NATIONAL STANDARDS FOR TOTAL SYSTEM BALANCE

AIR DISTRIBUTION • HYDRONIC SYSTEMS •
SOUND • VIBRATION •
FIELD SURVEYS FOR ENERGY AUDITS

Fourth Edition, 1982

●

ASSOCIATED AIR BALANCE COUNCIL

1133 15th Street N.W.
Washington, D.C. 20005

PUBLICATION INFORMATION

National Standards for Total System Balance is published by the Technical Information Division of the Associated Air Balance Council.

The Council is a non-profit, industry-supported association with membership composed of qualified agencies in the profession and business of testing and balancing of heating, ventilating, and air conditioning systems.

The AABC will not endorse any product, method of manufacture, or any item whatsoever for reasons of profit or otherwise.

The information compiled in this book is designed to establish a minimum set of field testing and balancing standards. The Standards will provide the Architect, Engineer, and Owner with a properly and totally balanced system.

Additional copies of these Standards may be obtained from National Headquarters.

FIRST EDITION 1967
SECOND EDITION 1972
THIRD EDITION 1979
FOURTH EDITION 1982

COPYRIGHT © ASSOCIATED AIR BALANCE COUNCIL 1982

All rights reserved. No part of this publication may be reproduced by photocopying recording, or by any other means, or stored, processed or transmitted in or by any computer or other systems without the prior written permission of the AABC.

ISBN 0-910289-00-X

Although great care has been taken in the compilation and publication of this volume, no warranties, express or implied, are given in connection herewith and no responsibility can be taken for any claims arising herewith.

Comments, criticisms, and suggestions regarding the subject matter are invited. Any errors or omissions in the data should be brought to the attention of AABC Headquarters.

PREFACE

The AABC National Standards, 1982 is the latest step to encourage and maintain professional performance in the industry. This publication revises and replaces all National Standards previously published by the Council.

This manual is a performance standard for the Total System Balancing industry.

The 1982 Edition contains 28 chapters. These are divided into five sections:

    I. General
   II. Instrumentation
  III. Design Standards for Total System Balance
  IV. Specifications
   V. Procedures

Section I—General—Provides information regarding AABC, certification standards, and warranty.

Section II—Instrumentation—Describes acceptable instruments and their approved uses, accuracy standards, and recommended calibration for verification procedures.

Section III—Design Standards for Total System Balance—Provides information for the Architect and Engineer regarding the minimum balancing devices required in a system for proper Testing, Adjusting and Balancing.

Section IV—Specifications—Provides the Architect, Engineer, and Owner with sample minimum specifications for all types of systems.

Section V—Procedures—Provides the Architect, Engineer, and Owner with the recommended procedures for a Total System Balance, including pre-construction plan checks and periodic mechanical construction review. Provides the Architect, Engineer, and Owner with a required certified report for **Total System Balance**. It also provides information for the analysis and summary of data, procedures for data verification, and their recommended uses.

DEDICATION
H. Taylor Kahoe

The late H. Taylor Kahoe is remembered by all who knew him as one of the greatest individuals ever associated with the industry.

"He enriched every life he ever touched." This is a good measure of the support and encouragement Taylor gave to his friends and associates.

His devotion to the establishment of the Test & Balance Industry and the Associated Air Balance Council did, in fact, enrich the lives of all its members. He inspired them to perpetuate the concept of the AABC and to uphold the standards for which he so diligently devoted his entire effort.

Taylor began his career as a mechanical coordinator and inspector for a prominent engineering firm. During this tenure his exposure to the improper performance of HVAC systems inspired him to search for a solution to guarantee proper system performance. He determined that proper performance is totally dependent upon total system balancing.

Through study, research and field application, Taylor devised a total system balance procedure forming the basis from which all current procedures have been developed.

As an active member of the ASHRAE committee on testing and balancing, Taylor wrote over 20 exceptionally fine technical articles for industry publications.

Since the development of this Standards Manual is made possible as a result of the pioneering efforts of H. Taylor Kahoe, it is only fitting and proper that this publication be dedicated to his cherished memory.

FOREWORD

This 1982 Edition of the AABC National Standards is entirely reorganized and rewritten. Care has been taken to maintain the position of presenting Standards for **Total System Balance.**

This new Edition represents the first time the entire scope of the **Total System Balance** profession has been compiled and presented in concise Standards form.

Section IV, SPECIFICATIONS, is designed to allow the Engineer and Architect to produce a comprehensive specification for a specific project without extensive research or writing. These specifications are in modular form so that the segments applicable to a specific project can be lifted out and assembled as the specification document.

Some of the new areas presented in these Standards are:
1. Specifications for Variable Air Volume Systems.
2. Design Standards for **Total System Balance** of both air and hydronics.
3. Standards for specifying and locating balancing devices in both air and hydronic systems.
4. Procedures for preconstruction plan checks and construction reviews for **Total System Balance.**
5. Approved methods of analyzing and verifying **Total System Balance** Reports.

All of the material in these Standards was reviewed at first, second, and final draft by the Standards Committee of AABC. The material was also submitted in draft and final form to the AABC Board of Directors.

TOTAL SYSTEM BALANCE

TOTAL SYSTEM BALANCE is the process of testing, adjusting and balancing environmental and other systems to produce the design objectives.

ACKNOWLEDGMENT

The material in this manual was compiled and produced under the direction of Lynn Wray, P.E., and Leo Meyer in cooperation with the AABC Standards Committee. This Committee has donated many days in designing the book outline and format and in reviewing manuscripts for technical accuracy. The members of the Standards Committee are:

Bernard S. Moltz, T.B.E. (Chairman)
Air Balance Corp. of Florida
Deerfield Beach, Florida

J. Michael Nix, T.B.E.
Delta-T, Inc.
Dallas, Texas

Theodore Cohen, T.B.E.
Air Conditioning Test and
   Balance Co., Inc.
Great Neck, New York

Kenneth M. Sufka
Executive Director
Associated Air Balance
   Council

Larry A. Johnson, T.B.E.
Thermal Balance, Inc.
Lexington, Kentucky

Acknowledgment is also given to all members of the Associated Air Balance Council for their generous support in making this publication possible.

Special acknowledgment is given to Mr. Arnold Pearl, who served as AABC's National Director during the Council's formative years.

TABLE OF CONTENTS
AABC STANDARDS MANUAL

SECTION I—GENERAL
    CHAPTER
      1—THE AABC CONCEPT
      2—THE AABC ORGANIZATION
      3—CERTIFICATION STANDARDS
      4—AGENCY QUALIFICATIONS
      5—WARRANTY

SECTION II—INSTRUMENTATION
      6—BASIC PHILOSOPHY OF MEASUREMENT
      7—TEMPERATURE MEASUREMENTS
      8—PRESSURE MEASUREMENTS
      9—VOLUME MEASUREMENTS
      10—ELECTRICAL MEASUREMENTS
      11—ROTATIONAL SPEED MEASUREMENTS
      12—SOUND MEASUREMENTS
      13—VIBRATION MEASUREMENTS

SECTION III—DESIGN STANDARDS FOR TOTAL SYSTEM BALANCE (TSB)
      14—BALANCING DEVICES
      15—SYSTEMS

SECTION IV—SPECIFICATIONS
      16—AABC GENERAL SPECIFICATIONS
      17—SUPPLY AIR SYSTEMS—GENERAL
      18—LOW PRESSURE AIR SYSTEMS
      19—MEDIUM AND HIGH PRESSURE AIR SYSTEMS
      20—VARIABLE AIR VOLUME SYSTEMS
      21—RETURN AND EXHAUST AIR SYSTEMS
      22—HYDRONIC SYSTEMS
      23—SPECIAL SYSTEMS
      24—TEMPERATURE CONTROL SYSTEMS

SECTION V—PROCEDURES
      25—PRECONSTRUCTION PLAN CHECK AND CONSTRUCTION REVIEW
      26—REPORTS AND REPORT FORMS
      27—REPORT ANALYSIS PROCEDURES
      28—REPORT VERIFICATION PROCEDURES

APPENDIX

CHAPTER 1

THE AABC CONCEPT

1.1 TOTAL SYSTEM BALANCE

Total System Balance is the process of testing, adjusting, and balancing environmental and other systems to produce the design objectives.

Air conditioning is defined by the American Society of Heating, Refrigerating and Air-Conditioning Engineers, Inc. as "The process of treating air to control simultaneously its temperature, humidity, cleanliness, and distribution to meet the comfort requirements of the conditioned space."

Each air treatment process in the conditioning system contributes a specific function to produce proper environmental conditions. However, it is the coordinated action of all of these processes in a system—each related to and influencing the others—that produces the desired conditions. If any one of these coordinated functions does not perform as designed, the final results will affect system performance. **Total System Balance** is the process of testing, adjusting, and balancing each system component so that the entire system produces the results for which it was designed. It is a science that requires proper use of instruments, evaluation of readings, and adjusting the system to design conditions. The mere ability to use an instrument does not qualify a person as a Test and Balance Engineer or Technician. Qualification requires training and years of field experience in applying proven techniques and in analyzing gathered data.

1.2 ASSOCIATED AIR BALANCE COUNCIL

The Associated Air Balance Council (AABC) was established in 1964 to meet the needs for training and to establish acceptable **Total System Balance** techniques and standards. The AABC has four major functions:

A. Promote training, develop and distribute technical data.
B. Establish National Standards for **Total System Balance.**
C. Conduct a certification program for testing and balancing agencies.
D. Advance the state of the art of the Industry.

1.3 GENERAL INFORMATION

The Associated Air Balance Council is a non-profit, industry-supported association. Membership is composed of certified agencies engaged in the profession and business of **Total System Balance.** The AABC symbol and name is copyrighted, and use thereof is prohibited without written consent from the Association. To maintain its established independence, the Council will not endorse any product or method of manufacturing.

Election of officers is held annually during the National Meeting. The President, Vice President, Secretary-Treasurer, and Board of Directors are elected by the membership.

Technical contributions for publication are made to the Council by the membership.

Membership and Certification in AABC is open to any qualified test and balance agency. The agency must qualify as an independent test and balance company, having no affiliation with manufacturers, installing contractors, or engineering firms. This criteria was established to assure unbiased performance without conflict of interests.

Chapter 1—The AABC Concept

Issuance of membership and certification will be made solely upon the applicant's ability to meet the requirements as stated in Chapter 3.

1.4 OBJECTIVES OF THE COUNCIL AND ITS CERTIFIED MEMBERS

All certified member agencies of AABC are responsible for upholding the following standards and practices established by the Council:

A. Provide the Engineer or client with completely reliable documentation of the system and its performance.
B. Provide system information that can be verified at all times by the Engineer or client.
C. Provide unbiased opinion of the deficiencies encountered and propose corrective action.
D. Provide the air conditioning industry with established, uniform test methods for rating field performance.
E. Encourage engineering and research development in the field of **Total System Balance.**
F. Collect, evaluate, and disseminate technical information from member agencies to improve methods of **Total System Balance.**
G. Cooperate with and assist all government and private agencies in serving the public's best interests in all matters directly concerned with **Total System Balance.**
H. Represent the test and balance industry in all matters involving methods, study, standards, and procedures.
I. Establish membership qualifications for agencies and personnel.
J. Encourage a standard of business ethics.
K. Enforce and maintain the member certification program.
L. Enforce compliance with the AABC National Standards, 1982.

1.5 TECHNICAL STANDARDS

The primary technical effort of the Association is to develop accurate and reliable testing procedures which can be used and adopted as testing and balancing standards by all authorized agencies.

The Association establishes technical standards for calibrating field instrumentation.

The Association develops training programs for Test and Balance Engineers and Technicians.

1.6 APPLICATION GUIDES

The Association publishes application data and guide manuals to provide the most current state of the art for **Total System Balance.**

A list of AABC technical publications is available from National Headquarters upon request.

CHAPTER 2

THE AABC ORGANIZATION

2.1 ORGANIZATION (Fig. 2.1)

Membership is available to qualified independent test and balance agencies worldwide.

Membership is by agency. Each Member Agency must have in its employ an AABC Certified Test and Balance Engineer. This person must meet the Standards described in Chapter 4, and is issued an Annual Certificate from National Headquarters in Washington, D.C.

Membership in the United States and Canada is divided into four geographical areas:

Zone 1: Eastern United States
Zone 2: Central United States
Zone 3: Western United States including Hawaii and Alaska
Zone 4: Canada

Agencies outside the United States and Canada are classified as Associate Members and are eligible to form a chapter to better serve their individual needs.

Each of the three zones in the United States has an AABC Vice President who is on the Board of Directors. The Canadian Chapter is represented on the Board of Directors.

The three zones in the United States are divided into nine regions. Each Region has a designated AABC Representative (see map, Fig. 2.1).

AABC is governed by a Board of Directors which is comprised of the elected officers of the organization. All officers serve terms as provided by the Bylaws.

AABC BOARD OF DIRECTORS AND OFFICERS

President
Executive Vice President
Secretary-Treasurer
Vice President—Eastern-Zone 1
Vice President—Central-Zone 2
Vice President—Western-Zone 3
Director-at-Large (Immediate Past President)
Director-Canadian Chapter—Zone 4

The President is also Chairman of the Board of Directors. The directives of the Board are implemented by an Executive Director and Staff at National Headquarters.

2.2 MEMBERSHIP DIRECTORY

An "International Certified Agency Directory" is published by AABC annually. This Directory provides names and addresses of:

- The Board of Directors and Officers
- All Member Agencies
- Regional Representatives
- Member Agency Personnel
- National Headquarters Staff
- General Information concerning AABC, and a copy of the "National Project Certification Performance Guaranty"

The current AABC INTERNATIONAL CERTIFIED AGENCY DIRECTORY is available from AABC National Headquarters.

2.2
Chapter 2—The AABC Organization

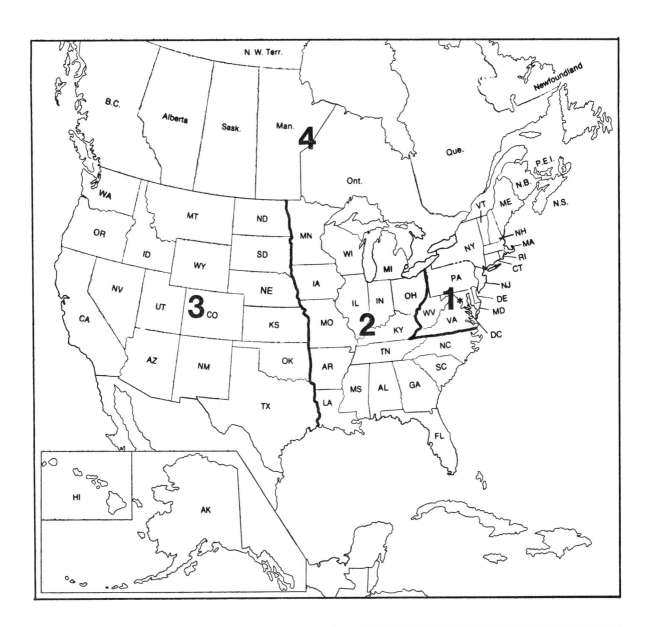

Fig. 2.1: AABC regions

ZONE 1	ZONE 2		ZONE 3		ZONE 4
Region 7 Delaware Maryland Pennsylvania Virginia West Virginia	**Region 4** Illinois Iowa Minnesota Missouri Wisconsin	**Region 6** Indiana Kentucky Michigan Ohio	**Region 1** Arizona California Hawaii Nevada	**Region 3** Colorado Kansas Nebraska New Mexico North Dakota Oklahoma South Dakota Texas Utah	Canada
Region 8 Connecticut Maine Massachusetts New Hampshire New Jersey New York Rhode Island Vermont	**Region 5** Alabama Arkansas Georgia Louisiana Mississippi North Carolina South Carolina Tennessee	**Region 9** Florida	**Region 2** Alaska Idaho Montana Oregon Washington Wyoming		

CHAPTER 3

CERTIFICATION STANDARDS

3.1 CERTIFICATION STANDARDS

All applicants for AABC Membership must:

A. Provide a written statement that the applicant is an independent agency having no affiliation with engineers, architects, installing contractors, or manufacturers of components of environmental systems. "Affilation" is defined to include financial interest or exchange of personnel.

B. Prove that the agency has performed testing and balancing for not less than three years immediately preceding application for membership.

C. Provide documented evidence of successful completion of fifteen Total System Balancing projects as a balancing agency—not as an individual endeavor.

D. Submit ten letters of endorsement from Engineers.

E. Provide a list of instruments owned by the agency, with manufacturer's name, model number and serial number where applicable. This list must meet or exceed the Standards for "Required Basic Instrumentation" as specified in Chapter 4 of the AABC National Standards, 1982.

F. Provide a written description of the agency operating facilities, and certify that these facilities meet or exceed the requirements specified in Chapter 4 of the AABC National Standards, 1982.

G. Provide a written resumé of the applicant describing and verifying experience in the field of Heating, Ventilating and Air Conditioning.

H. Prove in writing that properly trained field personnel and supervisors are being employed by the Agency. Give a brief resumé of the experience and education of each Technician engaged in **Total System Balance**.

I. Provide proof of financial stability of the agency (See Chapter 4).

3.2 MEMBERSHIP (Fig. 3.1)

Membership in the AABC is assurance to the client that work performed by the AABC Certified Agency will be completed to professional, quality standards. AABC membership provides for the following:

A. Each AABC Member is a Certified Agency that is pledged to perform all work in compliance with the AABC National Standards, 1982. Members are recertified on a yearly basis and reissued an annual AABC Certificate. (Fig. 3.1)

B. For each project, AABC Member Agencies are authorized to issue a National Project Certification Performance Guaranty in accordance with Chapter 5 of the AABC National Standards, 1982.

C. Each AABC Agency must have an AABC Certified Test and Balance Engineer in the organization. This Certified Test and Balance Engineer must meet the AABC National Standards, 1982 for education and experience and must also pass a rigid Certification Test that is administered by AABC National Headquarters. Each AABC Certified Test and Balance Engineer is assigned a Registration Number which is entered on each Test and Balance Report as an assurance of professional work.

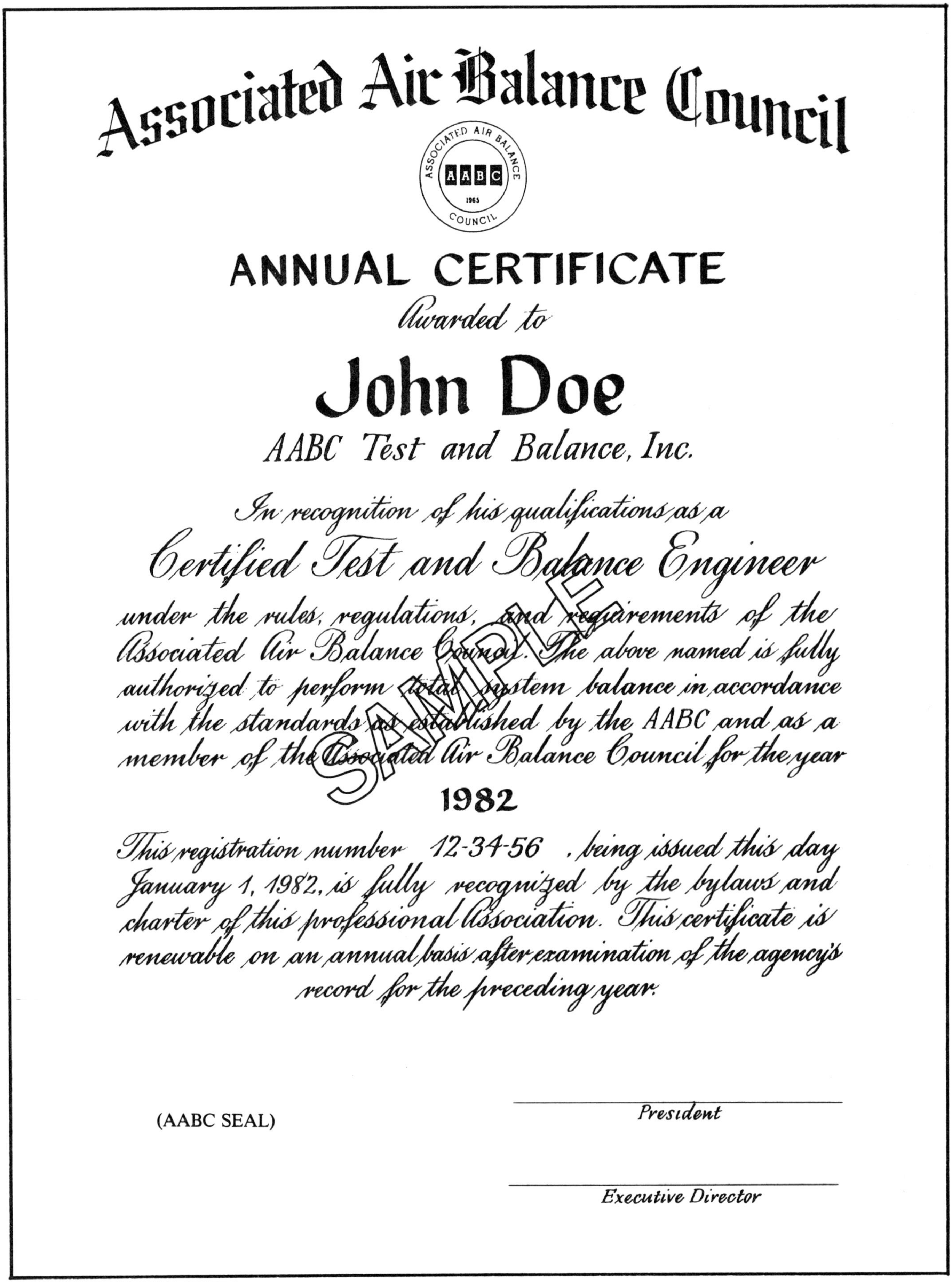

Chapter 3—Certification Standards

Associated Air Balance Council

ANNUAL CERTIFICATE

Awarded to

John Doe

AABC Test and Balance, Inc.

In recognition of his qualifications as a

Certified Test and Balance Engineer

under the rules, regulations, and requirements of the Associated Air Balance Council. The above named is fully authorized to perform total system balance in accordance with the standards as established by the AABC and as a member of the Associated Air Balance Council for the year

1982

This registration number 12-34-56, being issued this day January 1, 1982, is fully recognized by the bylaws and charter of this professional Association. This certificate is renewable on an annual basis after examination of the agency's record for the preceding year.

(AABC SEAL)

_____
President

_____
Executive Director

Fig. 3.1: Annual certificate

CHAPTER 4

AGENCY QUALIFICATIONS

4.1 GENERAL

The agency or company applying for membership and certification must have operated as an independent test and balance agency for a period of not less than three years immediately prior to application. During this three-year period, the agency must have continuously performed testing and balancing work as an agency—not as an individual. This time period is to provide the Council with a method of verifying the ability and experience of the applicant agency.

The applicant agency must have an established place of business, separate and distinct from a home or residence.

The applicant must furnish the Council with satisfactory proof of financial responsibility. The satisfactory proof shall be in the form of a certified financial statement. In lieu of this, a Cash Bond or Surety Bond in the minimum amount of $50,000.00 is required.

The applicant must furnish the Council with satisfactory proof of sole ownership and possession of the following complete sets of instruments. See Section II for detailed specifications of the instruments.

4.2 REQUIRED BASIC INSTRUMENTATION
(One required unless otherwise noted)

A. **HYDRONIC DIFFERENTIAL PRESSURE GAGES**
   0 to 50 Inches WG
   0 to 100 Inches WG
   0 to 30 Feet of Water (or greater)

B. **ANEMOMETERS**
   Rotating Vane
   Deflecting Vane

C. **TACHOMETER**
   Chronometric Type

D. **PITOT TUBES**
   0 to 18"
   0 to 24"
   0 to 36"
   0 to 48"
   0 to 60"

E. **ELECTRIC METERS**
   Portable Volt-Amp Meter
   Power Factor Meter

F. **PSYCHROMETERS**
   Sling or Powered

G. **FLOW HOOD**

H. **SMOKE SET**
   Gun
   Candles

I. **SOUND PRESSURE METER WITH OCTAVE BAND ANALYZER**

J. **THERMOMETERS**
   Glass Stem
   Dial
   Pyrometer
   Digital
   Recording

K. **MANOMETERS**
   0 to 10" Inclined and Vertical Scale
   0 to 1" Inclined Scale
   0 to 0.25" Inclined Scale (0.005" increments)
   18" U-Tube

L. **BOURDON TUBE GAGES**
   −30" Hg to 30 psi
   0 to 60 psi
   0 to 150 psi
   0 to 300 psi
   Test gage to verify other gages

M. **AIR DIFFERENTIAL PRESSURE GAGES (MAGNETIC LINKAGE)**
   0 to 0.5" WG
   0 to 1.0" WG
   0 to 2.0" WG
   0 to 4.0" WG
   0 to 8.0" WG

The applicant must furnish the Council with at least ten written endorsements by Engineers in the locality establishing qualifications and capabilities as a competent, independent Testing and Balancing Agency.

The applicant must have in its employ, on a full-time basis, at least one AABC Certified Test and Balance Engineer.

The applicant must maintain properly equipped and staffed facilities that are capable of compiling and distributing appropriate reports and data established from field measurements. These facilities must also be capable of furnishing the Engineer with completely reliable documentation of system information.

4.3 AABC CERTIFIED TEST AND BALANCE ENGINEER QUALIFICATIONS

The person who is certified by AABC as a Test and Balance Engineer must meet the following qualifications:

.1 Education

The applicant must have submitted a resumé of educational background which has been approved as satisfactory by AABC.

.2 Experience

Not less than ten years test and balance experience.

Five years of this must have been in continuous field experience in actual testing and balancing work. In addition, the applicant must pass the AABC examination for certification.

4.4 AABC TEST AND BALANCE TECHNICIAN QUALIFICATIONS

The **Technician** who is approved as an AABC Qualified Test and Balance **Technician** must meet one of the following requirements:

Five years experience in Testing and Balancing and pass an AABC qualifying test.

Completion of the AABC Testing and Balancing **Technician** Apprentice Program and pass the AABC qualifying test.

Certified AABC technicians must satisfactorily complete the Council's three-year Apprenticeship Training Course.

CHAPTER 5

WARRANTY

5.1 GENERAL

Member agencies of AABC are authorized by the Council to issue a "National Project Certification Performance Guaranty" (Warranty) on all work performed. This assures the client that his project will be balanced in accordance with plans and specifications and the AABC National Standards, 1982.

This chapter discusses the responsibilities of the Council in connection with the Warranty. It also outlines the enforcement procedures used wherein the warranty is invoked.

5.2 STATEMENT OF RESPONSIBILITY

The AABC Warranty explicitly states that:
A. The Certified Agency will test and balance all systems in accordance with the plans and specifications.
B. All systems will be balanced to optimum performance capabilities within the limitations of the design and installation.
C. If an Agency fails to comply with the specifications, the client may invoke the AABC Warranty. It is then the responsibility of the Associated Air Balance Council to provide supervisory personnel to assist the Test and Balance Agency to complete the project.
D. AABC shall not be required to accomplish **Total System Balance** if the system is improperly designed and/or installed; or if the equipment is found to be inadequate.

5.3 ENFORCEMENT PROCEDURES

A client may invoke the Warranty by filing a complaint against the Test and Balance Agency with the AABC National Headquarters.

The Executive Director will in turn begin a fact-finding mission by discussing the problem with all concerned parties. He shall also be responsible for keeping the AABC Board of Directors informed of the situation as it progresses.

If disagreement exists between the client and the Test and Balance Agency, an officer of the Association shall be assigned to arrange a meeting with the concerned parties. If necessary, the officer shall also conduct a personal investigation at the project site.

Based upon the findings, the investigating officer shall file a report along with recommendations to the AABC Executive Director and to an Executive Committee.

If the Test and Balance Agency is found to be at fault by the Executive Committee, immediate corrective action will be taken.

Ask for the AABC National Performance Guaranty Certificate on all Projects.

Chapter 5—Warranty

Associated Air Balance Council
Canadian Chapter
NATIONAL PROJECT CERTIFICATION
PERFORMANCE GUARANTY

Pursuant to the agreement between:

....
A.A.B.C. CERTIFIED TESTING & BALANCING FIRM

and ....
The certified testing and balancing Firm, will test and balance all systems in accordance with the plans and specifications
as prepared by ....
All systems as outlined in these specifications shall be balanced to optimum performance capabilities of the equipment and design. Testing shall be done in accordance with the standards as published by the ASSOCIATED AIR BALANCE COUNCIL.
If for any reason the Firm listed above fails to comply with the specifications, with the exception of termination of business by the Firm, equipment malfunction, inadequacy, or improper design, which prevent proper balancing of systems, the CANADIAN CHAPTER of the ASSOCIATED AIR BALANCE COUNCIL will provide supervisory personnel to assist the Firm to perform the work in accordance with AABC standards.
THE WORK COVERED BY THIS CERTIFICATE IS GUARANTEED FOR ONE YEAR FOLLOWING DATE OF REPORT
As part of this Performance Guarantee, the engineer, or architect may call upon the AABC Canadian Chapter to assist with all technical problems pertaining to the optimum balance of the systems. No extra charges for these services will be made by the above firm or by the AABC Canadian Chapter.

PROJECT ....
OWNER ....
ARCHITECT ....
CONSULTING ENGINEER ....
        CONTRACTOR ....
        REGISTRATION NUMBER ....
        DATE REGISTERED ....
        APPROVED BY ....
        AABC CERTIFIED FIRM ....

NATIONAL PROJECT CERTIFICATION
PERFORMANCE GUARANTY

Pursuant to the agreement between:

_____
AABC CERTIFIED TESTING & BALANCING AGENCY

and_____
the certified testing and balancing agency, will test and balance all systems in accordance with the plans and specifications
as prepared by:_____

    All systems as outlined in these specifications shall be balanced to optimum performance capabilities of the equipment and design. Testing and balancing shall be done in accordance with the standards as published by the Associated Air Balance Council.

    If for any reason, the Agency listed above, fails to comply with the specifications, with the exception of termination of business by the Agency, equipment malfunction, inadequacy, or improper design, which prevents proper balancing of systems, the Associated Air Balance Council will provide supervisory personnel to assist the Agency to perform the work in accordance with AABC Standards.

    As part of this Certifications Guaranty, the engineer, or architect may call upon AABC to assist him with all technical problems, or field problems pertaining to the final balanced condition of systems. No extra charges for these services will be made by the above agency or by AABC National Headquarters.

Project_____
Architect_____
Consulting Engineer_____
Copy Registered With National Headquarters
                  Date_____
T.B.E. No._____
         by_____
CERTIFIED AABC AGENCY

AABC Headquarters
1000 Vermont Avenue, N.W.
Washington, D.C. 20005

# CHAPTER 6

BASIC PHILOSOPHY OF MEASUREMENT

6.1 OVERVIEW

The necessity to maintain minimum standards for field measurement of air and hydronic systems has long been recognized by the Engineering Profession.

The Associated Air Balance Council, in the interest of producing accurate field measurements, sets forth these minimum standards for **Total System Balance** in this section. These methods are considered to be minimum requirements for all AABC Certified Members.

This chapter presents a standard for the Philosophy of Measurement.

6.2 TERMINOLOGY

For this standard the following definitions will be used.

- **Accuracy:** The capability of an instrument to indicate the true value of the measured quantity.
- **Calibration:** The process of adjusting an instrument to fix, reduce, or eliminate deviation.
- **Deviation:** The maximum departure between the calibration curve of an instrument and a straight line drawn to give the most favorable accuracy.
- **Error:** The difference between the indicated value and the true value of a measured quantity.
- **Flow Meter:** A device used for measuring fluid flow quantities. A flow meter generally requires a gage for indication.
- **Gage:** An instrument used to indicate differential pressure.
- **Human Error:** Inaccurate interpretation of a fluctuating indicated reading, incorrect or illegible recording of readings, inaccurate interpolations, parallax, or inaccurate conversion of units.
- **Instrument Error:** A quality defect in the instrument or in its original design from improper selection, poor maintenance and adjustment, or inadequate calibration.
- **Precision:** The repeatability of measurements of the same quantity under identical conditions.
- **Range:** A statement of the upper and lower value limits between which an instrument can be applied.
- **Reliability:** The probability that the repeatability and accuracy of an instrument will continue to fall within specified limits.
- **Repeatability:** See Precision.

6.3 GENERAL

When at the job site, the Test and Balance Agency shall have in its possession the required instrumentation to obtain proper measurements. Instruments shall be properly maintained and transported in such a manner as to provide protection against damage due to vibration, impact, moisture or any other condition that may render them inaccurate.

Instruments shall have been calibrated within a period of six months prior to starting the project. Proof of calibration shall be maintained with the instrument. Instruments shall be recalibrated upon completion of the work when required by the client to prove reliability.

6.4 SELECTION

Only instruments which have the maximum field measuring accuracy and are best suited to the function being measured shall be used.

6.5 APPLICATION

Instruments shall be applied as recommended by the manufacturer. If it is necessary to deviate from the recommendations due to field conditions, the modification shall be accomplished with logic and thought.

Scale range shall be proper for the value being measured. Instruments with minimum scale and maximum subdivisions are recommended.

6.6 TECHNIQUES OF MEASUREMENT

.1 Allowable Degree of Error

Measurements in the field shall be sufficiently free from error to match the task and allow proper judgments to be made. To obtain a degree of error-free data that is beyond reasonable necessity is not economically justifiable. Judgment by the Test and Balance Agency in these situations must be exercised. The Test and Balance Agency cannot be expected to obtain data with a more reasonable tolerance than the system permits. Further, error-free results cannot be expected if provisions for proper measurements were not included in the original design or installation.

.2 Quantity of Readings

When averaging values, a sufficient quantity of readings shall be taken which will result in a repeatability error of less than 5%. When measuring a single point, the reading shall be repeated until the same two consecutive values are obtained.

.3 Parallax

All readings shall be taken with the eye at the level of the indicated value to prevent parallax. Instruments with mirrored scale backing shall be used when practical.

.4 Fluctuating Readings

Pulsation dampeners shall be used where necessary to eliminate error involved in estimating average of rapidly fluctuating readings.

.5 Location of Reading

Measurements shall be taken in the system where best suited to the task. Seldom are ideal locations for optimum accuracy available in the field. Best judgment shall be exercised to obtain results in the most practical manner.

.6 System Installed Meters

Where meters for balancing are permanently installed in the system by others, the Test and Balance Agency cannot be expected to provide data with any less degree of error than the accuracy of the meter itself, its location, or the manner in which it is installed. If true values are to be measured, instruments with a sufficient degree of accuracy must be provided.

Two meetings of the AABC general membership are held each year where technical updates and presentations are given.

SECTION II—INSTRUMENTATION

CHAPTER 7

TEMPERATURE MEASUREMENTS

7.1 OVERVIEW

Temperatures will be measured in a system where necessary to properly perform the **Total System Balance.** If other temperature measurements are to be obtained, these requirements must be clearly defined in the specifications.

This chapter presents a standard for temperature measurement related to **Total System Balance.** Emphasis is on the proper use of instruments.

Temperature measurements are a check and not a primary balancing procedure. Balancing by temperature measurements is unreliable because of variable conditions such as atmospheric, load, pressure, temperature, and automatic control operations. Other conditions which cause error in obtaining true value measurements are stratification of the medium and position of the sensing element.

7.2 GENERAL

To prevent errors in temperature measurement, ample time shall be allowed to obtain temperature equilibrium.

When measuring temperature differential across any system component, the same instrument shall be used for all readings.

7.3 GLASS STEM THERMOMETERS

Mercury filled thermometers may be used for measuring temperatures in the range of −40°F to 500°F. Alcohol filled thermometers may be used for measuring temperatures in the range of −100°F to 250°F provided their accuracy is confirmed by a calibrated thermometer.

Glass stem thermometers shall be constructed with their scales etched into the glass. The error of the thermometers shall not exceed ± one scale division.

Glass stem thermometers are generally limited to hand-held immersion readings in applications where precise measurements are required. These thermometers can also be used as a standard for calibrating and verifying all other types of thermometers.

7.4 BIMETALLIC THERMOMETERS

Where bimetallic thermometers are used, the following conditions shall be met:

Thermometers shall be verified against a standard glass stem thermometer before use. If a bimetallic thermometer has received any type of severe shock or the dial has been twisted on the stem, it shall not be used before calibration. The range of the selected thermometer shall be such that the measured temperature is in the upper 50% of the scale.

7.5 DIGITAL THERMOMETERS

Where glass stem thermometers cannot be applied, a digital thermometer shall be used when precise readings are required for chilled water temperature, air temperature differential across a cooling coil, and setting minimum outside air quantities, etc.

The thermometer shall indicate the temperature in tenths of degrees and shall be accurate to ± 0.1 degree.

The thermometer shall be capable of field calibration to assure accuracy. When taking readings with a digital thermometer, sufficient time, as published by the manufacturer, shall be allowed to obtain maximum accuracy.

7.6 THERMOCOUPLE AND RESISTANCE THERMOMETERS

Fluid and surface temperatures may be measured with resistance or thermocouple thermometers.

Dial and pointer thermometers shall have a mirrored scale to allow readings to be taken without parallax.

Thermocouple thermometers shall not be used unless the temperature of the instrument is the same as the ambient.

7.7 PSYCHROMETERS

Sling or powered psychrometers shall be used for measuring wet bulb and dry bulb temperatures.

The wick on the wet bulb thermometer shall be wetted with distilled water, and the wick shall be kept clean.

When using the sling psychrometer, readings shall be taken until two repeatable, consecutive values are obtained.

7.8 RECORDING THERMOMETERS

Temperatures may be recorded by using recording thermometers when specifically required by the contract documents. In general, this procedure will not be required as a part of the **Total System Balance.**

7.9 THERMOMETER WELLS

Thermometer wells will be used for measuring fluid temperatures if they are provided by others and are correctly located. When using thermometer wells, the well must be filled with a thermal conducting material.

The permanently installed thermometers are not to be used for measuring temperatures for **Total System Balance** unless their accuracy is verified against a glass stem standard thermometer (see 7.3).

7.10 SURFACE TEMPERATURE MEASUREMENT

Although not a recommended procedure, surface temperature measurements of piping and equipment can be used for approximating the fluid temperature of hot water heating systems. It is recommended that thermometer wells or pressure-temperature taps be provided where any fluid temperatures are to be measured. Surface temperature reading should not be used for measuring fluids below 150°F.

When measuring surface temperatures, the surface to be measured must be cleaned and polished to a bright condition. Care shall be exercised to avoid error caused by sensing ambient temperature. Recommended procedures by the instrument manufacturer shall be used. Ample time must be allowed so that the temperature indicated has stabilized.

The Council maintains a full staff at AABC National Headquarters in Washington, D.C. Please contact this office for additional information or assistance.

CHAPTER 8

PRESSURE MEASUREMENTS

8.1 OVERVIEW

This chapter presents a standard for pressure measurements in air and hydronic systems for **Total System Balance.** The measurement of pressures is the most critical part of balancing the flow in any system.

Flow rate of a fluid is determined by the difference in pressure between two points in a system. Several different pressure measurements are a part of the standard procedure in the balancing process. Among these are: static pressure, velocity pressure, total pressure, and pressure differential across various components of the air conditioning system. For these measurements, it is essential that an accurate instrument with the correct range be applied properly.

Field measurement conditions cannot provide sufficient accuracy to be compared to manufacturers' published data, since field and laboratory conditions are different. Typical problem areas are:
A. Measuring pressure rise across a fan.
B. Measuring static pressure and pressure differential across various components of the systems.

Pressures must be measured in air and hydronic systems where needed for proper balancing to be performed. In addition, normal operating values under balanced conditions at various points in the systems must be measured. The results must be recorded in the Test and Balance Report for use by the Engineer and/or the Owner. Generally, readings will not be taken at other locations in the systems unless specified in the contract documents.

8.2 AIR PRESSURE MEASUREMENTS

.1 Locations

Static pressure measurements shall be taken at the locations in the air systems as indicated in **Section IV, Specifications.** After readings are observed and recorded, all test holes shall be securely and neatly plugged. Use of CaPlugs or equivalent is acceptable. Taping is not acceptable.

.2 Air Manometers

The basic instrument for field pressure measurements in an air system is the manometer. A manometer shall be used as a standard to verify accuracy of all other air pressure measuring instruments. Manometers shall be provided with a means of leveling, a means of making a zero adjustment, and a mirrored scale or other means to prevent parallax.

The types of manometers and their minimum application are:

A. Inclined manometer with 0.005 inch subdivisions shall be used for pitot tube traverses where air velocities are less than 1000 FPM. It may be used for higher velocity values if desired.
B. Inclined manometers with not larger than 0.01 inch subdivisions shall be used for pitot tube traverses where air velocities are in excess of 1000 FPM and less than 4000 FPM.
C. Vertical manometers with not larger than 0.1 inch subdivisions shall be used for pitot tube traverses where the air velocity exceeds 4000 FPM.

Readings shall be taken with the manometer only if the instrument is at ambient temperature, has a clean tube, is properly vented, leveled, and adjusted for zero at no pressure differential.

.3 Pitot Tubes (Fig. 8.1)

A standard pitot tube (Fig. 8.1) shall be used for determining velocity pressure in a duct. Smaller pitot tubes, geometrically similar to the standard tube may be used. The standard technique for its use is detailed in Chapter 9, "Volume Measurements."

.4 Air Differential Pressure Gages

A. Magnetically linked differential pressure gages (Magnehelic) may be used for reading static pressures in air systems. They are not to be used with pitot tubes to determine airflow quantities.
B. The gage range shall be selected so the reading is in the upper half of the scale.
C. These instruments are position sensitive. They must be placed in the same position when reading as when they were calibrated.
D. Magnetic gages shall be checked frequently against a standard manometer to ensure maximum accuracy.

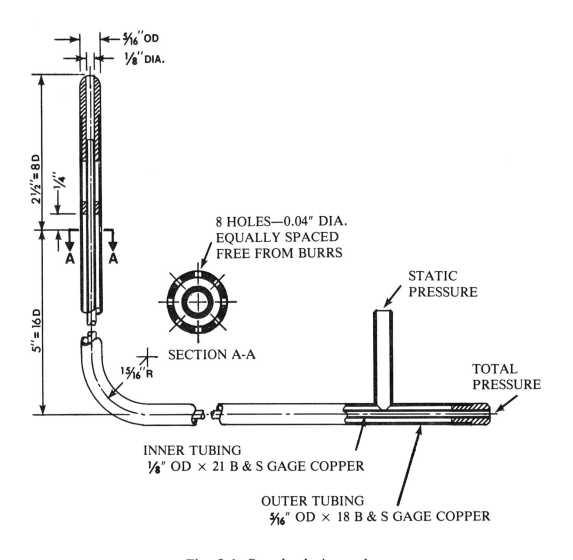

Fig. 8.1: Standard pitot tube

E. When measuring static pressure in an air system, a pitot tube or static pressure tip designed to eliminate the effect of velocity pressure shall be used. Where static pressure is relatively constant across the airstream (low velocity plenums, straight duct, etc.), a side wall tap with tubing attached to the meter may be used. The tap must be located so that the static pressure reading is unaffected by velocity pressure.

8.3 HYDRONIC PRESSURE MEASUREMENTS

.1 Locations

Static pressure measurements shall be taken at the locations in hydronic systems as indicated in Section IV, if pressure taps are provided. In addition, differential pressures across permanently installed meters shall be measured. The Test and Balance Agency shall not be required to tap into systems to obtain readings.

.2 Water Manometers

Mercury filled manometers may be used for reading differential pressure in hydronic systems. Mercury traps must be provided.

.3 Bourdon Tube Gage

A. The Bourdon Tube gage is the most commonly used gage for measuring static pressure in a hydronic system.
B. The gages used for **Total System Balance** shall be of industrial quality with a minimum $4^1/_2''$ dial. The range shall be proper for the pressure being measured. Accuracy shall be $^1/_2$ of 1% of the scale. Generally, the indicated pressure should be in the upper half of the scale.
C. Pulsations shall be dampened to stop rapid fluctuations in the readings.
D. When measuring hydronic differential pressures across any device with a Bourdon Tube gage, the same gage shall be used for all readings. The gage must be located at the same elevation for all readings to eliminate the effect of the weight of the fluid in the system at different heights.

.4 Test Gages

Each Test and Balance Agency shall own at least one test quality gage with proof of recent calibration. This is to be used for verifying the accuracy of the gages used in the field. The test gage shall be graduated in 0.5 psi subdivisions. The gage shall have a $4^1/_2''$ dial with a mirrored scale to prevent parallax and a knife edged pointer. Accuracy shall be $\pm^1/_4$ of 1% of the scale.

.5 Hydronic Differential Pressure Gages

A dual inlet pressure gage shall be used for measuring differential pressures across hydronic meters. The gages shall have a minimum 6" dial and shall be scaled in inches WG, feet of WG, or psi, as applicable.

The instrument range shall be selected so the reading is in the upper half of the scale. A gage with a sufficiently high range can be used for measuring static pressure differential across pumps.

CHAPTER 9

VOLUME MEASUREMENTS

9.1 OVERVIEW

This chapter covers the methods of measuring the rate of flow in both air and hydronic systems. In keeping with the AABC philosophy of **Total System Balance**, both types of flow measurements are treated equally.

Total System Balance is done in the field, not in the laboratory. Therefore, it is not always possible to apply the best and most accurate methods of measurement. The AABC Test and Balance Agency must use every means available to determine volume of flow. At times, several methods may be used as a crosscheck to determine the flow rates accurately.

The volume measurements described in this chapter for air and hydronic flow systems cover a wide range of measurement techniques. It is important to note that the accuracy of some of these methods is lower than others, but it may be necessary to use less accurate methods when no other ones can be applied. It is the obligation of the Test and Balance Agency to use every necessary means to achieve the desired results.

9.2 AIRFLOW PITOT TUBE TRAVERSES

.1 General

A. Whenever practical, pitot tube traverses shall be located at least $7^1/_2$ duct diameters downstream from an airflow disturbance or change in direction such as is caused by an elbow. The traverse shall also be $2^1/_2$ diameters upstream from such disturbances. Variations from the above shall be noted on the **Total System Balance** report.

B. Pitot tubes and manometers shall be in accordance with Chapter 8 of the AABC National Standards, 1982.

C. To measure Velocity Pressure (VP), the pitot tube shall be held parallel to the duct walls facing into the direction of airflow.

D. Air velocity shall be calculated using the following equation:

$$V = 4005 \sqrt{VP} \text{ (for Standard Air)}$$

V = Velocity in feet per minute
VP = Velocity pressure in inches of water

E. A correction factor for temperature and/or elevation shall be applied to the equation specified in Section 9.2.1D if air density varies from standard air by more than 10%.

F. Unless specified otherwise, pitot tube traverses shall be made in accordance with Section 9.2.2, Equal Area Procedure.

G. For calculations of Air Quantity (CFM):

Air Quantity = Average of Velocities in FPM* × Cross Sectional Area of the Duct in sq. ft.

H. Whenever a pitot tube traverse is taken, the static pressure in the duct at that point is to be measured and recorded in the **Total System Balance** report.

*Velocity Pressures cannot be averaged

Chapter 9—Volume Measurements

.2 Equal Area Procedure (Figs. 9.1, 9.2)

A. The Equal Area Procedure is recommended by the AABC as the preferred method for making pitot tube traverses.
B. Traverse for round duct, 6" diameter or smaller:
   1. The traverse shall consist of a total of 12 readings taken along two diameters at 90° to each other, and at centers of equal areas (Fig. 9.1).
   2. A pitot tube having a $1/8$" diameter shall be used.
C. Traverse for round duct larger than 6" diameter:
   1. The traverse shall consist of a total of 20 readings along two diameters at 90° to each other, at centers of equal area (Fig. 9.2).
   2. A pitot tube having a $3/8$" diameter shall be used.
D. Traverse for rectangular duct:
   1. At least 16 but not more than 64 readings shall be taken at centers of equal areas.
   2. If less than 64 readings are taken, the traverse points shall not be over 6" center-to-center.
   3. If 64 readings are taken, the traverse points may be over 6" center-to-center.
   4. A pitot tube having a $3/8$" diameter shall be used.

TUBE MARKINGS FROM WALL		
1	"Dia. × 0.043 =	"
2	"Dia. × 0.146 =	"
3	"Dia. × 0.296 =	"
4	"Dia. × 0.704 =	"
5	"Dia. × 0.854 =	"
6	"Dia. × 0.957 =	"

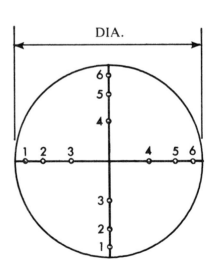

Fig. 9.1: Equal area traverse for round duct, 6" diameter or less

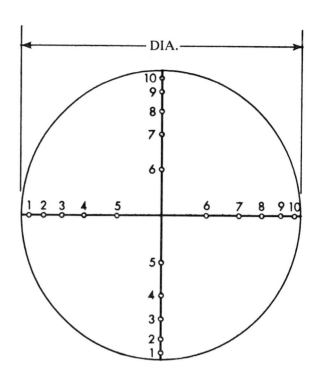

TUBE MARKINGS FROM WALL		
1	"Dia. × 0.026 =	"
2	"Dia. × 0.082 =	"
3	"Dia. × 0.146 =	"
4	"Dia. × 0.226 =	"
5	"Dia. × 0.342 =	"
6	"Dia. × 0.658 =	"
7	"Dia. × 0.774 =	"
8	"Dia. × 0.854 =	"
9	"Dia. × 0.918 =	"
10	"Dia. × 0.974 =	"

Fig. 9.2: Equal area traverse for round duct larger than 6" diameter

.3 Log Linear Procedure
(Figs. 9.3, 9.4, 9.5, 9.6, 9.7)

A. The Log Linear Procedure provides extreme accuracy ($\pm 3\%$). It considers the effect of friction along the walls of the duct.
B. The Log Linear Procedure may be used when specified.
C. Round duct traverse
   1. If the Log Linear Procedure is to be used, the three diameter, six-point method is the preferred traverse (Fig. 9.3).

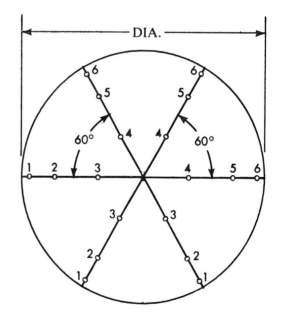

TUBE MARKINGS FROM WALL		
1	"Dia. × 0.032 =	"
2	"Dia. × 0.135 =	"
3	"Dia. × 0.321 =	"
4	"Dia. × 0.679 =	"
5	"Dia. × 0.865 =	"
6	"Dia. × 0.968 =	"

Fig. 9.3: Log linear, three diameter traverse

Chapter 9—Volume Measurements

2. If the three diameter method cannot be used (because of inaccessibility), then the two diameter method is acceptable. This method consists of two sets of ten readings, 90° apart (Fig. 9.4).

D. Rectangular Duct
   1. The minimum number of readings shall be 25.
   2. The points where the readings are to be taken shall be located at the intersection of the traverse lines as shown in Fig. 9.5.

This table indicates that any duct dimension that is less than 30" requires five traverse lines on that side. Thus a 28" × 20" duct will require 25 readings because each side would have five traverse lines.

A 32" × 20" duct will require 30 readings (6 traverse lines on the 32" side and 5 on the 20" side).

A 38" × 20" duct will require 35 readings (7 traverse lines on the 38" side and five on the 20" side).

3. The locations of the traverse lines shall be in accordance with Fig. 9.6. Fig. 9.7 gives an example of a 25 point traverse in a rectangular duct.

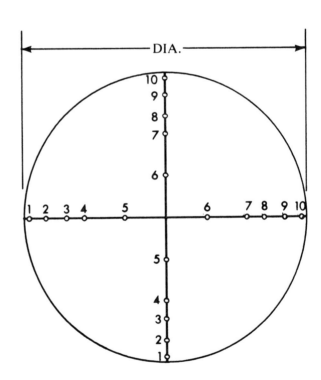

TUBE MARKINGS FROM WALL		
1	"Dia. × 0.019 =	"
2	"Dia. × 0.077 =	"
3	"Dia. × 0.153 =	"
4	"Dia. × 0.217 =	"
5	"Dia. × 0.361 =	"
6	"Dia. × 0.639 =	"
7	"Dia. × 0.783 =	"
8	"Dia. × 0.847 =	"
9	"Dia. × 0.923 =	"
10	"Dia. × 0.981 =	"

Fig. 9.4: Log linear, two diameter traverse

DUCT SIDE DIMENSION	NUMBER OF TRAVERSE LINES
Less than 30"	5
Over 30" and less than 36"	6
Over 36"	7

Fig. 9.5: Duct size and traverse lines

No. of Traverse Lines	Position Relative to Inner Wall						
5	0.074	0.288	0.5	0.712	0.926		
6	0.061	0.235	0.437	0.563	0.765	0.939	
7	0.053	0.203	0.366	0.5	0.634	0.797	0.947

Fig. 9.6: Location of traverse lines

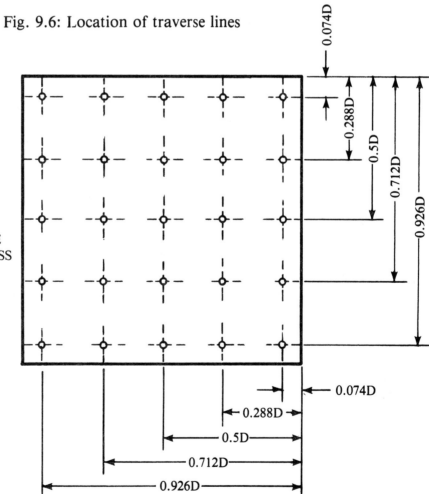

SAMPLE OF 25 POINT TRAVERSE (EACH SIDE OF THE DUCT IS LESS THAN 30")

NOTE: THE ABOVE EXAMPLE IS FOR A DUCT WITH BOTH SIDES LESS THAN 30".
A DUCT 32" × 20" WOULD REQUIRE 30 POINTS
(5 × 6 = 30)

Fig. 9.7: Example of a 25 point log linear traverse

9.3 ROTATING VANE ANEMOMETER

.1 General

The following are general specifications for the Rotating Vane Anemometer.

A. The minimum size shall be 4" diameter.
B. The dial face shall contain two small dials. One dial shall have 100 feet divisions, with one revolution equaling 1000 feet. The other dial shall have 1000 feet division, with one revolution equaling 10,000 feet.
C. The movement may be either ball-bearing or jeweled.
D. Electronic Vane or Mechanical Digital Anemometers are acceptable.

.2 Application

A. **The rotating vane anemometer shall not be used for measuring flow quantities through coils.** This is because the airstream exits in thin, high speed jets. Since the rotating vane anemometer is not an averaging instrument, it will provide an inaccurate reading.
B. The rotating vane anemometer shall not be used for measuring air velocities of less than 200 FPM unless the anemometer is designed especially for low velocity readings.
C. The airflow being measured must flow into the back of the anemometer and out the face.
D. All readings shall be taken with the anemometer handle in place to avoid interference to the airflow.
E. All readings shall be timed with a stop watch or a watch that indicates seconds.
F. Readings shall be taken for a minimum time of one minute.
G. When measuring airflow at a grille, the lowest velocity area shall be determined. The first reading in a series of at least four shall be from the area of the lowest velocity. The remaining readings should progress to the highest velocity areas.
H. Measurements shall consist of at least four consecutive readings of fifteen seconds each before disengaging the instrument. Sweeping action across the terminal face is not recommended.
I. Instrument manufacturer's correction factors for indicated velocities being measured shall be applied.
J. No correction factors for the anemometer are needed for temperature or altitude.

.3 Measuring Supply Grilles (Fig. 9.8)

A. Readings will be inaccurate beyond practical use if registers with restricted opposed blade dampers are measured with a rotating vane anemometer (registers are grilles with internal volume dampers). The airstream from such registers exits in thin, high speed jets. Since the rotating blade anemometer is not an averaging instrument, it will provide a high reading.
B. The rotating vane anemometer shall be held approximately 1" from the grille face to allow recovery from the vena contracta effect of the air passing between the face bars.
C. Face bars must be set at zero degrees deflection.
D. The area factor to be used for calculating CFM shall be from the inside dimensions of the grille frame (not the nominal grille size). See Fig. 9.8. This applies only if the instrument is held 1" from the face of the grille. Do not subtract the area of the face bars from the calculated area factor. If the smallest inside dimension of the grille is less than 4", use 4" (the diameter of the anemometer) for the calculations for the dimension.
E. The use of a rotating vane anemometer for proportionate balancing is satisfactory when properly applied. However, actual total flow quantities should be determined by a pitot tube traverse.

GRILLE—12" × 6" NOMINAL SIZE

$$A_K = \frac{11^5/_8" \times 5^5/_8"}{144" \text{ Per Sq. Ft.}}$$

$A_K = 0.45$ Sq. Ft.

Measured Velocity 475 FPM
CFM = (A_K × Corrected Velocity)
Instrument Correction Factor = 22FPM
Corrected Velocity = 475 − 22
     453

CFM = A_K × Corrected Velocity
    0.45 × 453 = 204

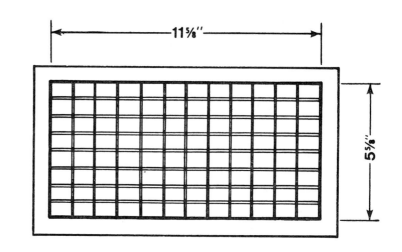

Fig. 9.8: Area factor for supply grilles is from inside frame dimensions

.4 Measuring Return and Exhaust Grilles (Figs. 9.9. 9.10)

A. When reading return and exhaust grilles, a shroud ring shall be attached to the dial-face side (Fig. 9.9). This shroud ring shall be held tightly against the face of the grille.

B. Airflow measuring techniques for return grilles shall be the same as Section 9.2.3.

C. The area for calculating air volume shall be by inside frame dimensions. If the grille has 45° fixed face bars, the dimension used for calculating area shall be that shown in Fig. 9.10.

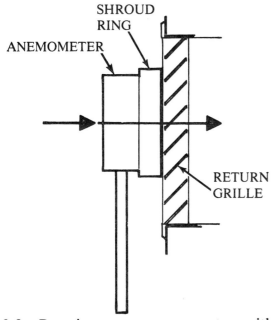

Fig. 9.9: Rotating vane anemometer with shroud ring

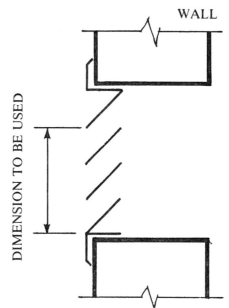

Fig. 9.10: Dimension to be used when calculating area factor for grilles with 45°, fixed, face bars

9.4 DEFLECTING VANE ANEMOMETERS

.1 General

A. The use of the deflecting vane anemometer for proportionate balancing is satisfactory when properly applied. However, actual total flow quantities should be determined by a pitot tube traverse.

B. The deflecting vane anemometer shall not be used with diffusers equipped with opposed-blade volume dampers in other than a fully open position.

C. The deflecting vane anemometer shall not be used for measuring grilles due to inconsistent velocities occurring at the face of the grille.

D. The deflecting vane anemometer is essentially restricted in use for measuring air velocity at supply diffusers.

.2 Application

A. The deflecting vane anemometer must be applied strictly in accordance with manufacturer's recommendations.

B. The deflecting vane anemometer is position sensitive. The scale must be held vertically.

C. The manufacturer's information regarding the position and angle of the probe; the number of readings to be taken, the area factor (A_K); and deflection pattern must be carefully followed.

D. The instrument manufacturer's correction factor for air density must be applied.

9.5 THERMAL ANEMOMETERS

.1 Heated Thermocouple Type

A. The heated thermocouple type thermal anemometer shall not be used in explosive atmospheres. The heated wire of the probe carries an electric current.

B. The probe must be exposed directly to the airstream when taking readings. It cannot be shielded. The probe of this instrument is extremely position sensitive.

C. The instrument manufacturer's instructions must be carefully followed.

D. The heated thermocouple type anemometer is calibrated for standard air—use a correction factor if the air density varies more than 5% from standard air conditions.

E. The heated thermocouple type may be used for velocities from 10 FPM to 2000 FPM.

.2 Hot Wire Type

The Hot Wire Thermal Anemometer:

A. Requires more frequent calibration than the thermal anemometer.

B. Can be used for low velocities.

C. Has accessories available to measure temperatures.

D. Is calibrated to standard air. A correction factor should be used if air density varies more than 5% from standard air conditions.

9.6 FLOW HOODS

A. Flow hoods equipped with direct reading instruments can be used to obtain reliable airflow quantities at terminals. Thus, use of these hoods reduces the time and effort required to measure airflow when compared with using instruments which require taking several readings and calculating the results by multiplying the average velocity by the effective area of the terminal. In addition, when the hood is properly applied, the readings will be unaffected by uneven face velocity or jet streams of air between the blades of integral dampers.

B. Flow hoods must be applied strictly in accordance with the manufacturer's instructions and all needed correction factors must be applied. The instrument on the hood must be calibrated for the position in which it will be held when the readings are taken. Hoods can be applied to low velocity outlets without the

SECTION II—INSTRUMENTATION

effect of back pressure when the manufacturer's recommendations are followed and corrections are made. Hood correction factors can be verified or confirmed in the field by pitot tube traverse.

C. Shop fabricated hoods whose readings have been verified against a known standard are reliable. A correction curve must be established for varying flow rates and actual field conditions.

D. Total air quantities of the system should be determined by a pitot tube traverse wherever practical.

9.7 OUTSIDE AIR QUANTITIES

.1 Overview

Generally, measuring outside air quantities is extremely difficult. Usually there is no good location to take a measurement, and reading quantities across louvers is unsatisfactory. In addition, Engineers now specify extremely small amounts of outside air to be handled by the systems. This creates a difficult task for the **Total System Balance** Agency.

The following methods for setting outside air quantities are recommended by the AABC when special provisions are not made for accurately measuring outside air quantities directly. "Special Provisions" are:

- Constant volume regulator at the outside air intake set to control at the required CFM.
- An accurate flow measuring station with balancing dampers in the fresh air intake.
- When an economizer cycle is used, a separate minimum outside air automatic damper, properly sized for the minimum air quantity shall be used. A damper sized for the maximum amount of outside air should not be used for minimum outside air quantity. It must be backed up by a constant volume regulator (PRV) or manual damper so the design amount of outside air can be set.

If a variable air volume system is involved, it may be well to consider the installation of a small outside air fan that is sized to handle the required quantity of minimum outside air to satisfy or overcome the fixed exhaust in the building along with the stack effect.

.2 Setting Mixed Air Quantities—Temperature Method (Fig. 9.11)

A. Mixed air quantities may be set by the temperature method using the same thermometer for all readings. A digital thermometer shall be used which complies with specifications, Chapter 7 of the AABC National Standards, 1982.

B. The mixed air quantities are determined by measuring the outside air temperature, the return air temperature, and fan discharge air temperature.

C. Temperatures shall be measured at a time when the difference between outside air and return air temperatures is the greatest.

D. Fan discharge air temperature shall be determined by taking a sufficient number of readings and averaging them to be sure that a true mixed air condition is being measured.

E. Fan discharge air temperature is measured because this is the location in the system where the outside air and return air have had the best opportunity to mix.

F. The air on the discharge side of the fan contains more energy than the air on the inlet side because work has been performed by the fan. For practical purposes, the temperature rise across the fan will equal approximately 0.5 times the measured total static pressure rise across the fan if the motor is outside the airstream. If the motor is inside the airstream, use 0.6.

G. Equation 1 shall be used to determine the percent of outside air when the outside air temperature, return air temperature,

9.10
Chapter 9—Volume Measurements

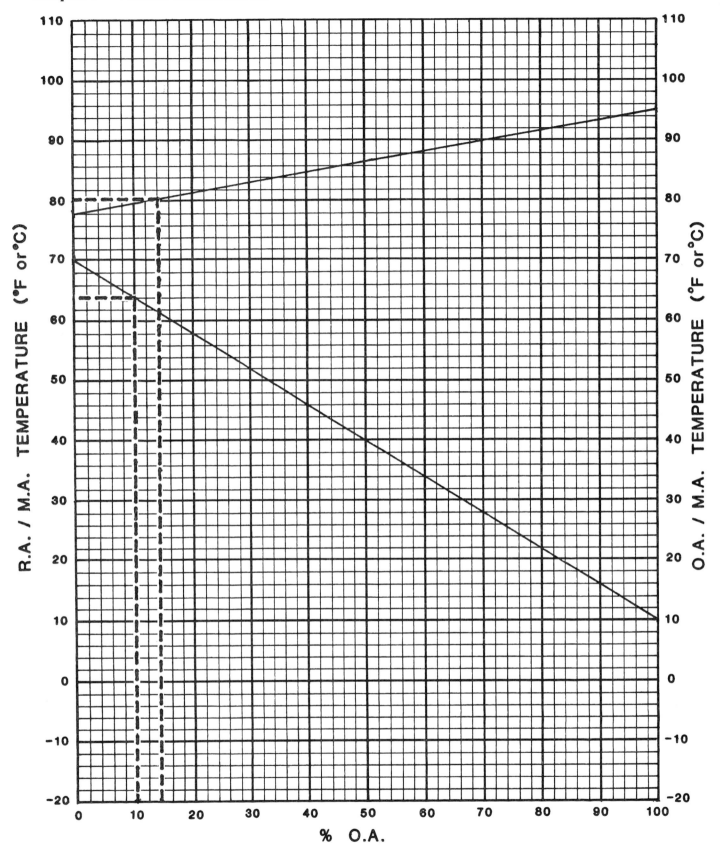

Fig. 9.11: % outside air vs. outside return and mixed air temperatures

average fan discharge air temperature and static pressure rise across the fan are known:

Equation 1:

$$\% \text{ OA} = \frac{T_R - [T_F - 0.5 \, (TSP)]}{(T_R - T_O)} \times 100$$

OA = Outside Air Quantity
T_O = Temperature Outside Air
T_R = Temperature Return Air
T_F = Average Temperature Fan Discharge
TSP = Total SP Rise Across Fan

H. Equation 2 shall be used to determine the necessary fan discharge air temperature when the required percent of outside air, return air temperature, outside air temperature, and static pressure rise across the fan are known:

Equation 2:

$$T_F = T_R - [\% \text{ OA} \, (\frac{T_R - T_O}{100})] + 0.5 \, (TSP)$$

I. The percent of outside air may be determined by the nomograph in Fig. 9.11 along with the use of the following equation:

Equation 3:
$$T_M = T_F - 0.5 \, (TSP)$$
T_M = Temperature Mixed Air

The nomograph is used as follows:
Plot the outside air temperature on the right-hand vertical axis, and plot the return air temperature on the left-hand vertical axis. Draw a line between these two points. Mixed air quantities may then be determined for any percentages by the intersection of the drawn line with the vertical and horizontal lines of the nomograph. The mixed air temperatures are read on the left-hand vertical axis, and the percent of outside air is read on the horizontal, bottom axis of the nomograph.

The lower diagonal line in Fig. 9.11 is an example of a plotted line for 70°F return air temperature and 10°F outside air temperature. The dashed lines show that if 10% outside air quantity is required, the temperature of the mixed air must be 64°F.

The top diagonal line is an example of a line plotted for 78°F return air temperature and 95°F outside air temperature. The dashed lines show that if the mixed air temperature is 80°F, the percent of outside air will be 14%.

9.8 AIR SYSTEM COMPONENTS
(Figs. 9.12, 9.13, 9.14)

.1 Coils

Use of heating and cooling coils in air systems to determine flow quantities by comparing manufacturer's ratings of flow to pressure differential is generally unsatisfactory. This method should be used for approximation only. This is because of the difference between the conditions in which the coil is tested in the field and the conditions under which the coil was rated. Some of the reasons are:
- Entering and leaving duct configurations.
- Nonuniform air velocities across the coil.
- Coil conditions such as cleanliness, bent fins, and corrosion.

.2 Induction Units

Nozzle pressure of induction units can be used to accurately determine the primary airflow through induction units. This is the recommended procedure.

.3 Variable Air Volume Terminals

Some variable air volume terminals have internal flow meters, such as orifice plates or center point total pressure and static tips. These cannot be used to determine flow quantities because the field conditions are not identical to the conditions under which they were rated. CFM quantities should be determined by summation of outlet quantities or pitot tube traverse.

9.9 HYDRONIC PITOT TUBE TRAVERSE
(Figs. 9.12, 9.13, 9.14, 9.15)

A. Water flow quantities in piping can be measured by a pitot tube and a differential pressure gage. For large piping, a pitot tube traverse is recommended.

B. Properly installed and located test stations must be provided in accordance with Chapter 14 of the AABC National Standards, 1982.

C. Flow quantities shall be calculated in accordance with Figs. 9.12, 9.13, and 9.14.

D. Pitot tube traverses are to be taken in the same manner as described for air in Chapter 9.

E. A double reverse pitot tube shall be used (Fig. 9.15) with a manometer or a differential pressure gage (see Chapter 8).

PIPE DATA				
PIPE		AREA SQ. FT.	GALLONS PER FOOT OF PIPE	I.D. INCHES
SIZE	SCH			
6	40	0.2006	1.50	6.065
8	20	0.3601	2.69	8.125
8	30	0.3553	2.66	8.071
8	40	0.3474	2.60	7.981
10	20	0.5731	4.29	10.25
10	30	0.5603	4.21	10.136
10	40	0.5475	4.09	10.02
12	20	0.8185	6.12	12.25
12	30	0.7972	5.97	12.09
12	40	0.7773	5.81	11.938
14	20	0.9758	7.30	13.376
14	30	0.9575	7.17	13.25
14	40	0.9394	7.02	13.124
16	20	1.290	9.66	15.376
16	30	1.268	9.49	15.25
16	40	1.2272	9.20	15.0
18	20	1.647	12.35	17.376
18	30	1.599	12.00	17.124
18	40	1.5533	11.64	16.876
20	20	2.021	15.15	19.25
20	30	1.969	14.75	19.0
20	40	1.9305	14.47	18.814

Fig 9.12: Pipe Data

VP	VEL	VP	VEL
1″	107.5 FPM	26″	547 FPM
2″	152 FPM	27″	588 FPM
3″	186 FPM	28″	568 FPM
4″	215 FPM	29″	578 FPM
5″	240 FPM	30″	588 FPM
6″	264 FPM	31″	597 FPM
7″	284 FPM	32″	607 FPM
8″	304 FPM	33″	617 FPM
9″	322 FPM	34″	627 FPM
10″	340 FPM	35″	635 FPM
11″	356 FPM	36″	645 FPM
12″	372 FPM	37″	654 FPM
13″	387 FPM	38″	662 FPM
14″	402 FPM	39″	671 FPM
15″	417 FPM	40″	679 FPM
16″	431 FPM	41″	687 FPM
17″	444 FPM	42″	696 FPM
18″	456 FPM	43″	705 FPM
19″	468 FPM	44″	712 FPM
20″	482 FPM	45″	721 FPM
21″	492 FPM	46″	730 FPM
22″	505 FPM	47″	736 FPM
23″	516 FPM	48″	744 FPM
24″	527 FPM	49″	753 FPM
25″	537 FPM	50″	758 FPM

Fig. 9.13: Velocity (FPM) vs. velocity pressure (inches of H_2O) for water flow, using combined reversed pitot tube

SECTION II—INSTRUMENTATION

TRAVERSE POINT	DIA. × DIM.	OUTER LOCATION	HORIZONTAL VP"	VELOCITY FPM	VERTICAL VP"	VELOCITY FPM
1	" ID × 0.026 =	"				
2	" ID × 0.082 =	"				
3	" ID × 0.146 =	"				
4	" ID × 0.226 =	"				
5	" ID × 0.342 =	"				
6	" ID × 0.658 =	"				
7	" ID × 0.774 =	"				
8	" ID × 0.854 =	"				
9	" ID × 0.918 =	"				
10	" ID × 0.974 =	"				
			TOTAL		TOTAL	
CENTER	" ID × 0.500					

$$\text{AVERAGE VEL.} = \frac{\text{TOTAL HORIZONTAL} + \text{TOTAL VERTICAL}}{\text{NUMBER OF TRAVERSE POINTS}}$$

MEASURED GPM = AVERAGE VEL. × (7.481 × AREA)

Fig. 9.14: Calculating flow quantities from hydronic pitot tube traverses

Fig. 9.15: Double reverse pitot tube

9.10 FLOW MEASUREMENT BY HYDRONIC METERS

.1 Overview

There are two meters commonly used in Environmental Hydronic Systems:
A. Venturi flow meters
B. Orifice flow meters

Both have an error of 1%, if properly applied.

Meters must be installed with the following considerations:

A. All meters in horizontal piping must be installed with their gage ports above the horizontal axis of the pipe.
B. Since corrosion of gage ports and orifices are a problem, proper system water treatment must be maintained.
C. Systems must be cleaned initially and maintained clean to prevent fouling the meters.

.2 Venturi Flow Meters

A. Venturi flow meters are the most commonly used because:
- They have a low pressure loss, since a carefully formed flow path provides very little dynamic loss. The pressure loss across the Venturi is approximately 25% of the indicated pressure differential. This is a very small amount because the pressure differential is very small.
- The accuracy of the Venturi is less affected by inlet and outlet conditions.
- The Venturi is less subject to fouling by debris and therefore provides consistently accurate readings.

B. **The Engineer should specify that the Venturi to be provided should have a ratio between throat and pipe diameters so that the pressure differential reading across the Venturi is between 25" to 50" WG.**

C. The differential pressure shall be measured with a differential pressure gage in accordance with Chapter 8 of the AABC National Standards, 1982.

.3 Orifice Flow Meters

A. The Engineer may elect to require the installation of orifice flow meters in piping systems because of their low initial cost. Those meters have a high degree of inherent accuracy when properly installed. The degree of accuracy generally increases with increase in pressure differential across the meter. However, consideration must be given to the need to provide pumps with sufficient head to overcome the system resistance as well as the dynamic loss across the flow meter.

B. Special care must be taken with regard to upstream and downstream flow conditions when using orifice flow meters. Piping requirements should be as shown in Figure 17, Diagrams G and H of Chapter 13, 1981 ASHRAE Fundamentals Handbook.

C. The differential pressure across these meters shall be measured with a diffential pressure gage in accordance with Chapter 8 of these AABC National Standards, 1982. If the orifice is to be used for future readings, special care must be taken to maintain the system in a clean state (see Overview, Section 9.10.1).

9.11 AVERAGING PITOT TUBE

A. The averaging pitot tube is often used in lieu of the Venturi or orifice flow meters because of lower initial cost. It is built with a single-ported static pressure tip and a multi-ported total pressure tip. The concept is to average the total pressure in one plane.

B. The averaging pitot tube is subject to inaccuracy of readings due to disturbances in the entering or leaving piping.

C. To achieve accuracy, the averaging pitot tube must be installed strictly according to manufacturer's instructions.

D. The piping size where the averaging pitot tube is installed may have to be reduced in order to raise the velocity pressure to ensure an accurate reading.

9.12 CALIBRATED BALANCE VALVE

A. The calibrated balance valve is a variable orifice flow meter.

B. The manufacturer's recommendations for minimum distance upstream and downstream of disturbances must be carefully followed.

C. A calibrated balance valve is not as accurate as a separate flow meter.

D. In many cases the pressure drop across a calibrated balance valve exceeds that of other types of valves used for balancing, such as full-ported plug valves, ball valves, etc.

E. Proper sizing of a calibrated balancing valve is very important. If the valve is too small, the pressure drop will be excessive, resulting in a need for a pump with a high head characteristic and, consequently, unnecessary energy consumption. On the other hand, if the valve is too large, inaccurate readings will result. The Engineer should control the size of any valves where the Contractor is allowed to substitute the manufacturer in order to ensure that it will be compatible with the system.

9.13 SELF-CONTAINED AUTOMATIC FLOW LIMITING DEVICE

See Chapter 22, Section 22.7 of the AABC National Standards, 1982.

9.14 FLOW MEASUREMENT BY PUMP DIFFERENTIAL HEAD (Fig. 9.16)

A. The pump can be used as an indicator of flow quantities. However, this technique will result in an estimate of flow—not an accurate reading.
B. Determining flow quantities by the pump will be an estimate because:
1. **The pump is not a meter.**
2. The pump curve will not represent the flow characteristics of the particular pump being used. The curve is general for that model, and characteristics may vary with each pump.
3. Field test conditions will be different from the laboratory rating conditions.
C. If the pump is not cavitating, and if it is free of air, the flow can be estimated by determining the static pressure rise across the pump and adding the calculated difference in velocity head that will result from differences in suction and discharge connection sizes. This expression of total head can be plotted on the pump curve to determine the **approximate** GPM.
D. To determine total dynamic head the following equation shall be used:

TDH = (Discharge SP − Inlet SP) + (Discharge VP − Inlet VP)

The static pressure rise can be read by use of a differential pressure gage or by a single Bourdon Tube Gage. If a Bourdon Tube Gage is used, the same one shall be used for both static pressure readings, and the gage must be held at the same elevation for both readings or a correction must be made for the difference in height.

To determine the velocity pressure, use the chart in Fig. 9.16.

Example: To determine the velocity pressure for a 2500 GPM pump with an inlet connection of 10″ and an outlet connection of 8″, start at 2500 GPM (Point A on the top of the chart). Read down to point B (the intersection of the 8″ diameter pipe line) and then read left to the velocity pressure scale. The velocity pressure is 5.5 feet WG. (Point C).

The velocity pressure for the 10″ pipe is found in the same manner. Start at point A and draw a line down to the 10″ diameter line (point D) then read across to the left at point E, which shows a velocity pressure of 2.2 feet WG.

9.15 FLOW BY HYDRONIC SYSTEM COMPONENTS

A. Estimates of flow can be made by calculation using the field tested pressure drop across a component as compared to the manufacturer's rated pressure drop. This method will provide only estimated flow quantities, because of the difference in field conditions and in laboratory conditions which established the manufacturer's data.

If manufacturer's ratings are **calculated** rather than derived from

9.16
Chapter 9—Volume Measurements

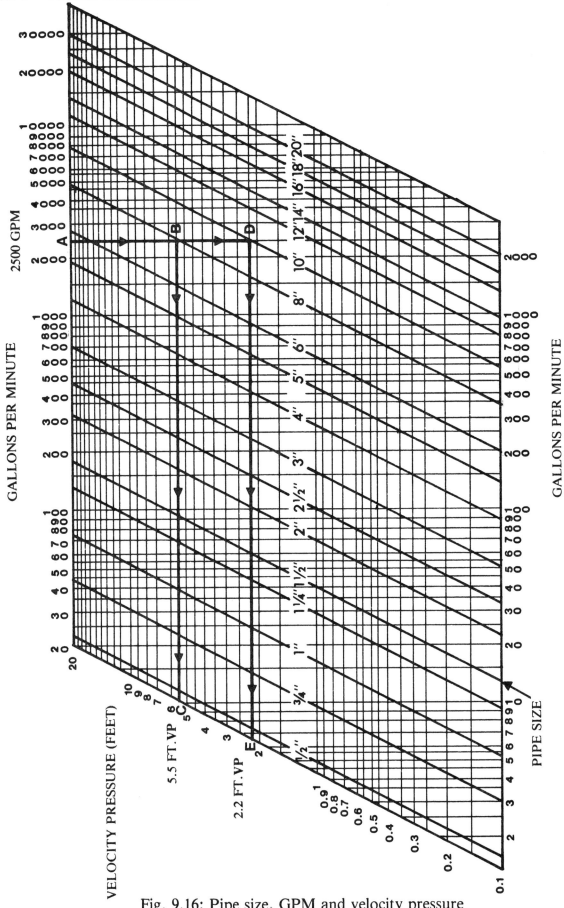

Fig. 9.16: Pipe size, GPM and velocity pressure

laboratory tests, the flow quantities derived by this method can only be regarded as rough estimates.

B. The accuracy of this method depends upon:
- The accuracy of the manufacturer's information.
- The accuracy of the field readings for pressure differential.

> The Council maintains a full staff at AABC National Headquarters in Washington, D.C. Please contact this office for additional information or assistance.

CHAPTER 10

ELECTRICAL MEASUREMENTS

10.1 OVERVIEW

This chapter presents a standard for the electrical measurements that are required for **Total System Balance**.

To date, most electrical measurements have been made to prevent overloading motors. The results of these measurements are not relevant to the driven equipment manufacturers' published data (expressed in brake horsepower) which were derived from measurements under laboratory conditions.

In the balancing procedure, the power input to the motor can be measured. The brake horsepower of the driven equipment can be estimated if the motor and drive efficiency are known. To complete the calculations requires the measurement of power factor, volts, and amperes (kilowatt input). Motor efficiency and drive efficiency must be known. If this information is required of the Test and Balance Agency, Engineers must require manufacturers of motors to submit certified performance data for each motor to be tested. Further, the Engineer should require the drive manufacturers to submit power load data for various combinations of sheaves which provide required speed ratios of a given unit.

10.2 GENERAL

Accurate, repeatable electrical measurements must be taken as a part of **Total System Balance** to ensure that powered equipment is operating in a safe non-overloaded condition. Normal operating voltage and amperage of all equipment involved in the **Total System Balance** work shall be measured.

10.3 INSTRUMENTS

.1 Volt-Ammeter

A. A hand-held portable multimeter capable of measuring both current and voltage shall be used for electrical measurements in the field.
B. The meter to be used shall have an accuracy of ±2% of the true value.
C. The meter shall have a digital display which indicates the measured value in tenths.
D. A clamp-on volt-ammeter with graduated scale and movable pointer may be used in lieu of a digital meter for field measurements provided the following conditions are met:
 1. Accuracy of the meter shall be not less than ±3% of the full scale.
 2. Instrument accuracy is carefully verified against a standard or a digital meter as described above.
 3. Readings shall be taken with the lowest possible scale range.
 4. The eye of the instrument reader shall be at the same level as the pointer to prevent parallax.

.2 Measuring Current and Voltage

A. When reading current, all phases shall be measured. Care shall be exercised to center the conductor in the jaw of the meter to avoid picking up any induced current from other sources or any control circuits.
B. When reading voltage, the potential between each phase shall be measured. Voltage measurements shall be made as close to the energized piece of equipment as practical.

Chapter 10—Electrical Measurements

.3 Power Factor Meter

A. When specifically called for, power factors shall be measured at selected items of energized equipment.

B. Power factor measurements shall be taken with a hand held clamp-on power factor meter having an accuracy of ±2% of true value.

C. When reading power factors, all phases shall be measured. In addition, the corresponding current of each phase and voltage between each phase shall be measured.

D. An approved alternative procedure is to measure power demand with a watt meter and compare it with measured volt-amperes using the following equations:

Single Phase

$$\text{Power factor} = \frac{\text{Watts}}{\text{Volts} \times \text{Amps}}$$

Three phase

$$\text{Power factor} = \frac{\text{Watts}}{\text{Ave. Volts} \times \text{Ave. Amps} \times 1.732}$$

10.4 BRAKE HORSEPOWER DETERMINATION

.1 BHP By Calculation

A. It is difficult to calculate Brake Horsepower accurately from field test data. To do so with minimum error, the following facts must be known:

$$\text{Average Amperes} = \frac{A_1 + A_2 + A_3}{3}$$

$$\text{Average Volts} = \frac{V_1 + V_2 + V_3}{3}$$

$$\text{Average Power Factor} = \frac{(PF_1 \times A_1) + (PF_2 \times A_2) + (PF_3 \times A_3)}{A_1 + A_2 + A_3}$$

$$A = \text{Amperes}$$
$$PF = \text{Power Factor}$$
$$V = \text{Volts}$$

B. For equipment driven by three-phase motors the following equation shall be used:

$$BHP = \frac{1.732 \times \text{Ave. A} \times \text{Ave. V} \times \text{Ave. PF} \times \text{Motor } E_{ff} \times \text{Drive } E_{ff}}{746}$$

$$
\begin{aligned}
A &= \text{Amperes} \\
\text{Ave.} &= \text{Average} \\
BHP &= \text{Brake Horsepower} \\
E_{ff} &= \text{Efficiency} \\
PF &= \text{Power Factor} \\
V &= \text{Volts}
\end{aligned}
$$

Motor Efficiency:
　Obtained from manufacturer for specific size and type at percent of loading.

Drive Efficiency:
　Obtained from manufacturer.

.2 BHP By Approximation

A. Brake Horsepower can be approximated from field data. To do so for three-phase motors, the following items must be known:
  1. From motor nameplate:
     a. Nameplate HP
     b. Nameplate Amperes
     c. Nameplate Volts
  2. Motor Full Load Efficiency × Power Factor (calculated by the following equation)

$$FLE_{ff} \times PF = \frac{NPHP \times 746}{1.732 \times NPA \times NPV}$$

$$
\begin{aligned}
FLE_{ff} &= \text{Full Load Efficiency} \\
PF &= \text{Power Factor} \\
NPHP &= \text{Nameplate Horsepower} \\
NPA &= \text{Nameplate Amperes} \\
NPV &= \text{Nameplate Volts}
\end{aligned}
$$

  3. Average Running Amperes $= \dfrac{A_1 + A_2 + A_3}{3}$

  4. Average Running Volts $= \dfrac{V_1 + V_2 + V_3}{3}$

5. Percent Efficiency × Power Factor at less than full load. This can be determined for three-phase motors from the following Table of "L" Factors (Fig. 10.1). These Factors were established by averaging the published information of several motor manufacturers.
6. Approximate Percentage of Load. This can be determined by the following equation:

$$\% \text{ Load} = \frac{(RA - 0.5\, NPA)}{(0.5\, NPA)} \times \frac{RV}{NPV}$$

RA = Running Amperes
NPA = Nameplate Amperes
RV = Running Volts
NPV = Nameplate Volts

B. Use the following equation to approximate Brake Horsepower:

$$BHP = \frac{1.732 \times RA \times RV\, (FLE_{ff} \times PF)\, L}{746}$$

BHP = Brake Horsepower
RA = Running Amperes
RV = Running Volts
$FLE_{ff} \times PF$ = Motor Full Load Efficiency times Power Factor
L = Percent Efficiency times Power Factor at less than full load

"L" Factors

Motor H.P.	Load 50%	Load 75%	Load 100%
1	0.710	0.885	1.0
1.5	0.715	0.890	1.0
2	0.780	0.935	1.0
3	0.795	0.940	1.0
5	0.835	0.950	1.0
7.5	0.875	0.970	1.0
10	0.890	0.975	1.0
15 thru 30	0.910	0.980	1.0
40 thru 125	0.940	0.990	1.0
150 thru 300	0.920	0.985	1.0

"L" Factor = Percent Efficiency × Power Factor

Fig. 10.1: Table of "L" factors

CHAPTER 11

ROTATIONAL SPEED MEASUREMENTS

11.1 OVERVIEW

This chapter presents a standard for rotational speed measurements for **Total System Balance.** Rotational speed measurements are among the simplest, most direct, and most accurate measurements to make. The following material describes the typical instruments required for such measurements and explains their uses. Also included is information on the measurement of fan RPM and special situations.

11.2 GENERAL

The final RPM of all belt driven equipment shall be accurately measured, and the data shall be recorded in the Total System Balance Report. In addition, the initial RPM of all equipment shall be recorded in the Report whenever speed changes are made.

Generally, measuring the RPM of motors, or equipment directly connected to constant speed motors, is not a standard requirement.

11.3 CHRONOMETRIC TACHOMETER

A chronometric tachometer shall be used for measuring rotational speed whenever the shaft of the item being tested is accessible. A chronometric tachometer is a hand-held instrument which is a combination of a precision stop watch and a revolution counter. It is the least position-sensitive of all available contact type tachometers. The accuracy of the instrument to be used shall be not less than $1/4$ of 1% of the dial scale.

For measuring speeds of less than 2000 RPM, an instrument which records 0 to 1000 RPM shall be used. For higher RPM, an instrument shall be used which records 0 to 10,000 RPM.

To ensure the most accurate reading, the instrument shaft must be held parallel to the rotating shaft. The correct tip for the shaft type must be used. It shall be in contact with the shaft for a sufficient time to reach speed before starting the measurement. The shaft shall be prepared to ensure proper contact without tip slippage.

When the shaft of a belt-driven item of equipment is not accessible, but the belt is, and the pitch diameter of the driven sheave is known, the RPM can be calculated by measuring belt speed. The equation to be used is:

$$RPM = \frac{IS \times 1.91}{pd + 0.2}$$

IS = Indicated Speed which is the indicated value on the dial when using a 2″ diameter attachment

pd = Pitch diameter in inches of the driven sheave

11.4 DIGITAL TACHOMETER

A digital tachometer having an accuracy of ± one digit may be used for measuring rotational speed. If used, extreme care must be taken to hold the instrument as nearly parallel to the rotating shaft as possible. In any event, readings shall be repeated until two consecutive, repeatable values are obtained.

11.5 STROBOSCOPE

A stroboscope may be used for measuring rotational speed when the shaft or belt is not accessible.

The stroboscope has an electronically controlled flashing light which can be manually adjusted to equal the frequency of the rotating part so it will appear to be motionless. Care must be taken to avoid reading harmonics of the actual RPM. This can be avoided by starting all readings at the low end of the instrument scale.

Any stroboscope used for **Total System Balance** work shall have an accuracy of ±1% of the scale being used.

11.6 PHOTO TACHOMETER

A photo tachometer may be used for measuring rotational speed when the shaft or belt is not accessible.

The photo tachometer is equipped with a light source and a photo electric cell which when pointed at a reflective strip placed on a rotating body can convert the signal into an indication of RPM.

To ensure that the reading to be taken will be in the upper half of the scale, a sufficient quantity of equally spaced markers should be placed on the sheave or rotating part.

Rotational speed is obtained by use of the following equation:

$$RPM = \frac{\text{Scale Reading}}{\text{No. of Markers Used}}$$

Any photo tachometer used for **Total System Balance** work shall have an accuracy of ±1% of the scale being used.

11.7 SPECIAL SITUATIONS

The AABC National Standards, 1982 does not require RPM to be measured when, in the judgment of the Test and Balance Agency, the above procedures cannot be practically applied.

When measuring RPM with any instrument other than a chronometric tachometer, special note shall be made in the report form.

11.8 MEASURING FAN RPM

When measuring fan RPM, the system must be in the same condition as when the balancing was performed.

All access doors or panels must be closed. If this is not possible, temporary access covers must be provided so the system resistance is the same as actual operating conditions.

An AABC newsletter is published monthly by the Council.

CHAPTER 12

SOUND MEASUREMENTS

12.1 OVERVIEW

The AABC Test and Balance Agency is responsible for the measurement of sound pressure levels in accordance with the contract specifications. This chapter presents the Standards for sound measurements for **Total System Balance.** The emphasis is on the instruments used in measuring sound pressure levels and the testing procedures for accurate sound level measurements. Also included are brief definitions of basic sound terminology and a general discussion of octave bands, combining decibels, acceptable sound pressure levels and noise levels.

12.2 TERMINOLOGY

For this Standard, the following definitions will be used.
- **Sound**: Results from a vibrating solid. Pressure waves are created (generally in air or liquid) by a vibrating solid. The impression interpreted by the ear to these pressure waves is sound.
- **Noise**: Unwanted sound.
- **Frequency**: Time rate of repetition of pressure waves. Sound frequency is stated in cycles per second (Hertz). One cycle per second is 1 Hertz (Hz). The audible frequency range is 20Hz to 20,000 Hz.
- **Octave**: A frequency range (band) whose upper limit is twice the frequency range of the lower limit.
- **Microbar**: Pressure equal to one millionth of one **bar.** A bar equals pressure at atmosphere.
- **Decibel (dB)**: Unit of measurement for sound pressure. Zero decibels has been established as the threshold of hearing for a person having excellent hearing at a frequency of 1000 Hz. Technically, zero dB equals 0.0002 Microbar.
- **Sound Pressure Level (SPL)**: Sound pressure level stated in dB.
- **Sound Power Level**: A rating of sound power (in dB) generated by equipment. This value is used by the Engineer to predict final sound pressure that will exist in a space. Sound power level cannot be field measured.
- **Noise Criteria Curve**: A curve representing how the human ear hears and reacts to a sound pressure level and its corresponding frequency.
- **N. C. Curve Chart**: A chart made of several Noise Criteria Curves ranging from values of NC 20 to NC 70.

12.3 GENERAL
.1 Octave Bands (Fig. 12.1)

Octave bands for this Standard are to be those stated in ASHRAE 1981 FUNDAMENTALS, page 7.3, Table 3. See Fig. 12.1 of this chapter.

Frequency Limits

Octave Band	(Hz)	Center Frequency
1	44-88	63
2	88-177	125
3	177-355	250
4	355-710	500
5	710-1420	1000
6	1420-2840	2000
7	2840-5680	4000
8	5680-11360	8000

Fig. 12.1: Octave Bands

.2 Adding or Subtracting Decibels (Fig. 12.2)

Decibels cannot be added or subtracted arithmetically. For this Standard, decibels are to be combined using the method described in ASHRAE 1981 FUNDAMENTALS, page 7.3, Table 5. See Fig. 12.2 of this chapter.

Difference Between Two Levels to be Combined, dB	0 to 1	2 to 4	5 to 9	10 and more
Number of decibels to be added to higher level to obtain combined level	3	2	1	0

Fig. 12.2: Combining two sound levels

(Note: These are for *unoccupied* spaces, with all systems operating.)

Type of Area	Recommended NC Criteria Range*
1. Private residences	25 to 30
2. Apartments	30 to 35
3. Hotels/motels	
a. Individual rooms or suites	30 to 35
b. Meeting/banquet rooms	30 to 35
c. Halls, corridors, lobbies	35 to 40
d. Service/support areas	40 to 45
4. Offices	
a. Executive	25 to 30
b. Conference rooms	25 to 30
c. Private	30 to 35
d. Open-plan areas	35 to 40
e. Computer/business machine areas	40 to 45
f. Public circulation	40 to 45
5. Hospitals and clinics	
a. Private rooms	25 to 30
b. Wards	30 to 35
c. Operating rooms	25 to 30
d. Laboratories	30 to 35
e. Corridors	30 to 35
f. Public areas	35 to 40
6. Churches	25 to 30**
7. Schools	
a. Lecture and classrooms	25 to 30
b. Open-plan classrooms	30 to 35**
8. Libraries	30 to 35
9. Concert halls	**
10. Legitimate theaters	**
11. Recording studios	**
12. Movie theaters	30 to 35

*Design goals can be increased by 5 dB when dictated by budget constraints or when noise intrusion from other sources represents a limiting condition.
**An acoustical expert should be consulted for guidance on these critical spaces.

Fig. 12.3: Recommended indoor design criteria for air conditioning system sound control

.3 Acceptable Sound Levels (Fig. 12.3)

For this Standard, acceptable ranges of sound levels for different types of spaces shall be those given in ASHRAE 1980 SYSTEMS HANDBOOK, page 35.16, Table 23. See Fig. 12.3 of this chapter.

.4 Noise Level (Fig. 12.4)

For this Standard, acceptable noise levels will be determined as specified in the N.C. Curve Chart, ASHRAE 1981 FUNDAMENTALS HANDBOOK; page 7.10, Fig. 10. See Fig. 12.4 of this chapter.

12.4 INSTRUMENTS

.1 Sound Meter

A. The Sound Testing Meter shall be a portable Type 1 General Purpose Testing Meter and shall conform to ANSI Standards S1.4—latest edition.

B. The sound testing meter and the calibrator shall be from the same manufacturer, and the calibrator must be the one specified for the sound testing meter microphone.

C. If recalibration adjustments greater than ±2 dB are required, the sound testing meter and the calibrator shall be returned to the manufacturer for inspection and recalibration.

.2 Microphones

A. The acceptable types of microphones used with sound testing meters are the Piezo-electric and the condenser microphones.

B. The microphone shall be capable of measuring up to 20 KHZ.

C. The manufacturer's directions shall be followed regarding the basic orientation of the microphone with respect to wave direction.

D. Microphones shall be protected from high humidity which may affect the diaphragm.

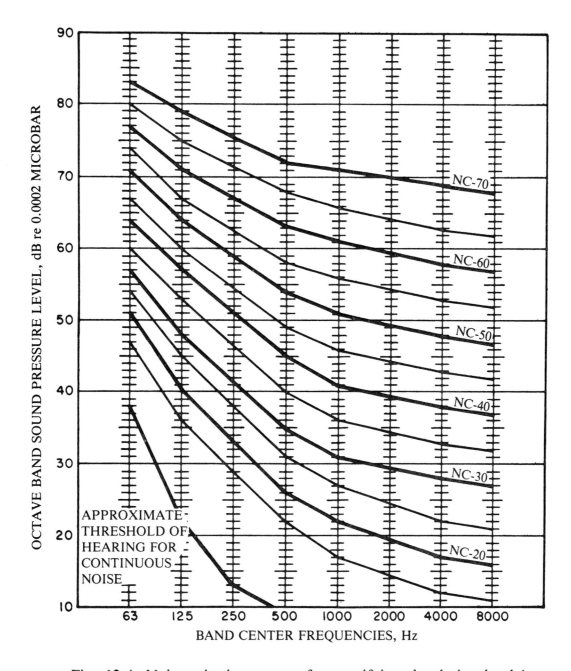

Fig. 12.4: Noise criterion curves for specifying the design level in terms of the maximum permissible sound pressure level for each frequency band

E. The microphones shall not be exposed to temperatures above 100°F or below −10°F. Refer to the manufacturer's directions for microphone use under conditions other than the above temperatures.
F. For all exterior readings, a wind screen shall be used on the microphone whenever wind velocities are greater than 5 miles per hour.

.3 Octave Band Analyzers

Octave Band Analyzers shall conform to ANSI Standard S1.11 "American Standard Specification for Octave, Half Octave and Third Octave Band Filter Sets."

12.5 TESTING PROCEDURE

Sound testing shall be performed according to the following Standards:
A. The Sound Testing Meter shall be calibrated before each use.
B. During calibration, a barometric pressure reading shall be taken and an appropriate calibration correction shall be made.
C. The operator shall clear the area under test of all persons except test personnel so unrelated disturbances will not be recorded either in background sound or in final readings.
D. The operator shall close all windows and doors.
E. All spaces where readings are to be taken shall be furnished in a normal manner. All items affecting room absorption should be in place (drapes, furniture, carpeting, etc.).
F. Tests shall be performed when the building is not occupied and outside noise levels are at minimum.
G. The Sound Testing Meter must be removed from its case since the case itself may interfere with the sound field at the microphone.
H. The operator shall stand well clear of the line of the sound source and the microphone.
I. If the meter is hand held, the microphone shall not be pointed directly at the sound source.
J. Sound Testing Meter readings in a room shall be taken at a height above the floor as recommended by the instrument manufacturer.
K. Sound pressure measurements should be taken with the equipment on and off. If the difference between the background reading and the reading with the equipment running is 10 dB or more, the background effect shall be considered as insignificant. If the difference is less than 7 dB, subtract the background effect from the total measured sound pressure level to determine mechanical equipment decibels. See Section 12.3.2.

All certified member agencies of AABC are independent and therefore have no affiliation with any manufacturer or installer of equipment.

CHAPTER 13

VIBRATION MEASUREMENTS

13.1 OVERVIEW

All environmental systems which serve a building produce vibrations which are transferred as sound through the air or as vibrations through components of the structure. Vibration may injure, tire, bother, or inhibit people. Related to equipment and structures, vibration may promote mechanical failure, poor performance, and excessive wear.

This chapter presents the Standards for measurement of vibrations during **Total System Balance**.

The AABC Test and Balance Agency shall measure the vibration of components of an environmental system where required by the project specifications.

13.2 TERMINOLOGY

- **Vibration:** Rhythmic movement of a solid to alternate sides of the position of equilibrium. Vibration has two basic dimensions with which the AABC Test and Balance Agency is concerned: frequency and deflection (also called amplitude).
- **Frequency:** The time rate of repetition of a vibration cycle. Frequency may be stated in CPM (cycles per minute) or in CPS (cycles per second).
- **Deflection:** The total movement of the vibration from one side to the other. Deflection is measured in mils (1/1000 of an inch).
- **Velocity:** The time rate of deflection. Velocity is proportional to deflection and frequency.

13.3 GENERAL

The AABC Test and Balance Agency shall measure and record vibration for deflection, and velocity if, and as required, in job specifications.

13.4 INSTRUMENTATION

.1 Vibration Testing Meter

The AABC Test and Balance Agency shall provide a portable vibration meter to measure vibration deflection and velocity as specified. The meter shall meet the following criteria:

A. It shall contain solid state circuitry.
B. It shall be equipped with a straight probe, vibration pickup and at least 12 feet of cable.
C. In the Velocity Mode it shall have a sensitivity of 0.1 in./sec. at a frequency range of 3 Hz to 1 KHz.
D. In the Deflection Mode it shall have a sensitivity of 0.001 inches at a frequency range of 3 Hz to 850 Hz.

.2 Vibration Calibrator

The operator shall use a vibration calibrator to field calibrate the testing equipment before and after each measurement session.

13.5 TESTING PROCEDURES

To complete the Vibration Measurements, the AABC Test and Balance Agency shall:
- Ensure that there is no construction in progress at the time of vibration testing.
- Ensure that all building construction equipment is turned off during the vibration testing session.
- Ensure that all vibration sources (traffic, elevators, etc.) other than environmental

system vibration be eliminated while testing is in progress.
- Place all testing instruments in positions independent of the equipment being tested.
- Clean the test area of grease or oil that could cause slippage of the vibration pickup.
- Clear the testing area of all persons other than test personnel.
- Make all required calibrations according to procedures specified by the instrument manufacturer before measurements are taken.
- Mount the vibration measuring device so that it does not interfere with the vibrating body.
- Position the measuring device according to the manufacturer's instructions to achieve maximum sensitivity.
- Measure each piece of operating equipment under actual operating conditions.
- Record all necessary vibration data on an AABC Certified Report Form and submit it to the Owner's Representative.

The Council maintains a full staff at AABC National Headquarters in Washington, D.C. Please contact this office for additional information or assistance.

Chapter 14

BALANCING DEVICES

14.1 OVERVIEW

This chapter deals with the minimum standards for balancing devices in air and hydronic systems. The material in this chapter is an essential part of the AABC committment to quality, professional **Total System Balancing.** The balancing procedure consists of imposing resistance in the system by use of dampers or valves to achieve the design objectives.

In theory the Engineer could design the system so that no field adjustment is necessary. However, this assumes that the calculations, and installations will be perfect. It also assumes that resistance of all system components will be exactly the same as Engineer's design data. Since this is not practical, balancing is needed.

Proper design and installation of balancing devices are vital to the effective performance of an air conditioning system. The Test and Balance procedure can be significantly improved with balancing devices that are properly selected and located to match the requirements of the air or hydronic distribution system. High quality instrumentation and skilled measurement techniques cannot offset a missing or improperly located volume control or measuring station. Inadequate selection and design of the balancing devices limits the effectiveness of system Test and Balance. It can also result in improper system operation and excessive use of energy. The objective of this chapter is to assist the Engineer in the design of the system so that it can be balanced.

14.2 AIR SYSTEMS

.1 Balancing Devices

A. The proper selection, location, and type of airflow control devices permit maximum air distribution in the system. These devices also allow equalizing the pressure drops in different airflow paths within the same system.
B. Balancing devices should be so located that they do not add to the sound levels generated by the mechanical equipment.
C. **Proper installation and location of balancing devices in the system eliminate the need for volume controls at grilles and diffusers.**

.2 Dampers (Fig. 14.1)

A. High pressure dampers should be constructed in accordance with Fig. 14.1.
B. Specify that mixing box dampers shall not have a leakage rate in excess of 3% of design CFM at design static pressure.

Chapter 14—Balancing Devices

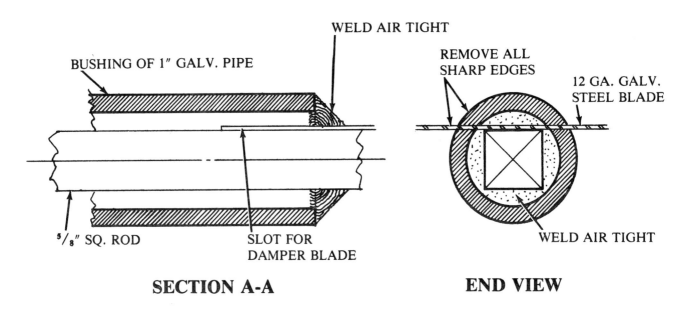

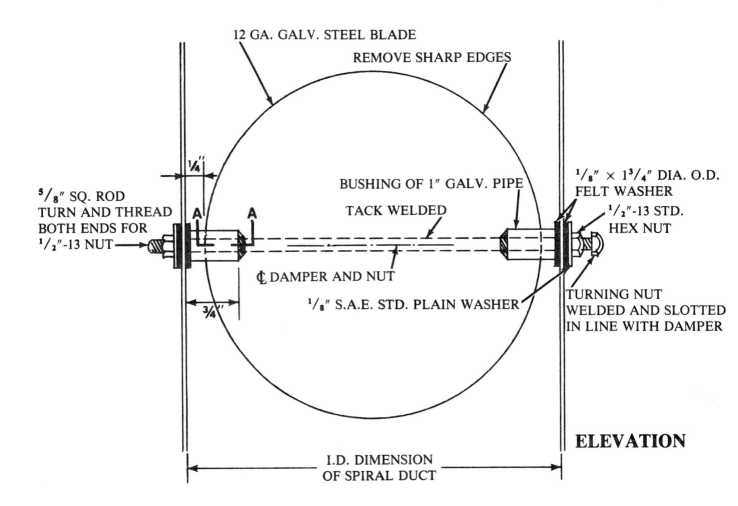

Fig. 14.1: Details for a high pressure duct damper

SECTION III—DESIGN STANDARDS FOR TOTAL SYSTEM BALANCE (TSB)

C. **A splitter is not a damper. It is a diverter and should not be used as a volume control device.** When used as a diverter, it must have at least one push rod with a locking device—not a quadrant handle. See SMACNA Duct Design Manuals for details. Diverters should be long enough to divert the airflow in the duct. As a minimum, the length of the diverter should be twice the size of the smallest throat of the nested connection.
D. Extractors are not balancing devices, and should not be used for this purpose.
E. Low pressure duct dampers must be adjustable, with locking quadrant handles.
F. The construction of all volume dampers shall be in accordance with the operating pressure of the duct system in which they are to be installed. Wherever possible, dampers should be multi-blade with opposed blade action. Each damper shall be adjustable, with locking quadrant handle, that has sufficient strength and rigidity for the pressures being controlled. For small round duct, a butterfly damper is satisfactory. For large round duct, a multi-blade damper should be used. The damper blades on any damper should not have formed edges as these interfere with airflow.

.3 Damper Location

Volume dampers are used for different purposes throughout the air distribution system. Following are some of the locations where volume dampers are recommended.
A. For supply, return, and exhaust systems, volume dampers must be located in each main branch duct. Each grille or diffuser connection must have a damper, but not as part of the outlet assembly. The damper in these sub-branches should be located as close to the main duct as possible.
B. Each reheat coil must have a volume damper, but far enough downstream so that the damper, when set in a restricted position, will not interfere with uniform airflow across the coil.
C. Every zone duct of a multizone system must have a volume damper.

.4 Supply, Return, Fresh Air Intake, and Relief Air Connections

Almost every air conditioning system has automatic dampers to control the mixture of fresh air, return air, and relief airflow rates. The operation of these dampers is controlled by temperature requirements of the system, not by airflow. It is important that manual volume dampers be installed in the fresh air intakes, relief air, and the return air connection between the return air fan and the mixed air plenum. The purpose of these volume controls is to balance the pressure drops in the various flow paths so that the pressure drop in the entire system will not change as the proportions of return air and fresh air vary to satisfy the temperature requirements.

14.3 HYDRONIC SYSTEMS
.1 Overview

Hydronic systems must have balancing devices that perform functions comparable to balancing devices in an air system—both are fluid flow systems. However, proper design for balancing devices is even more critical in hydronic systems. After the air system is operating, balancing devices can be added—even though it is costly. However, once the hydronic system is filled and operating, it cannot be entered readily to add balancing devices such as meters and valves. During system design, consult the AABC Test and Balance Agency for details and for locations of balancing devices needed in the system.

Flow meters must be installed in hydronic systems to measure flow rates. In the absence of proper measuring stations, it is possible to balance hydronic systems by use of pressure drops across coils, heat exchangers, and the pressure rise across pumps. However, this method is an approximation. The calculation of flow rates from these pressure differentials is based on manufacturers' data that may have been calculated or laboratory tested. In most cases the specific unit installed in the system that is being balanced has not been laboratory tested. In addition, the installation in the field may not conform to laboratory test conditions.

.2 General (Figs. 14.2, 14.3)

A. In general, provisions should be made for the use of gages, thermometers, and pressure taps in accordance with the chart in Fig. 14.2.
B. Pump gages shall be piped in accordance with Fig. 14.3.

.3 Pressure Taps

A. Pressure taps must be located as close as possible to coils or other equipment being measured.
B. When using hydronic meters, pressure taps must be on the side or the top of the pipe—never on the bottom. Taps on the bottom of the pipe will clog because they accumulate debris.
C. Where pitot tube traverses must be performed, specify two properly sized tapped openings 90° apart. For each tap, provide full-ported gate valves with plugs.

.4 Valves

A. A balancing valve must be provided at each terminal.

Point of Information	Manifold Gage	Single Gage	Thermometer	Test Well	Pressure Tap
Pump—Suction, Discharge	X				
Strainer—In, Out					X
Cooler—In, Out		X	X		
Condensers—In, Out		X	X		
Concentrator—In, Out		X	X		
Absorber—In, Out		X	X		
Tower Cell—In, Out				X	X
Heat Exchanger—In, Out	X		X		
Coil—In, Out				X	X
Coil Bank—In, Out		X	X		
Booster Coil—In, Out					X
Cool Panel—In, Out					X
Heat Panel—In, Out				X	X
Unit Heater—In, Out					X
Induction—In, Out					X
Fan Coil—In, Out					X
Water Boiler—In, Out			X		
3-Way Valve—All Ports					X
Zone Return Main			X		
Bridge—In, Out			X		
Water Makeup		X			
Expansion Tank		X			
Strainer Pump					X
Strainer Main	X				
Zone 3-Way—All Ports				X	X

Fig. 14.2: Balancing device requirements

SECTION III—DESIGN STANDARDS FOR TOTAL SYSTEM BALANCE (TSB)

B. A balancing valve must be provided at each return branch that serves more than one terminal.
C. In systems with multiple coils in a bank, there must be a balancing valve for each coil.
D. Do not use valves with a handle that has a limited number of holes or notches to lock it into a setting. Only use valves that allow infinite setting positions.

provide constant flow, a balancing valve must be installed in the by-pass piping. This will permit imposing resistance similar to the terminal so the circuit flow will remain essentially constant during modulation of the three-way valve (Fig. 14.5).

If a delegated secondary or tertiary pump is installed to provide constant circulation, no by-pass balancing valve is required.

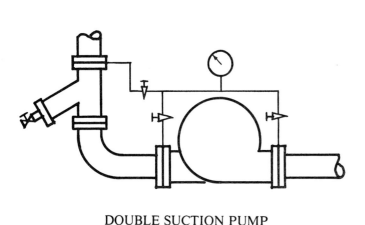
DOUBLE SUCTION PUMP

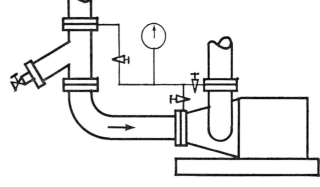
END SUCTION PUMP

Fig. 14.3: Pump gage piping

E. Specify that any balancing valve over 6" pipe size must have a worm gear and wheel for making settings.
F. All balancing valves should have memory devices which allow the valve to be returned to the balanced position after being closed.
G. Butterfly valves are not recommended for balancing. However, if they are used for balancing they should be carefully sized so that they have good throttling characteristics. Oversized valves are difficult to throttle.
H. Three-way automatic control valves, which allow bypass of terminals, are sometimes installed to provide constant circulation of systems. When a delegated pump is not installed at the terminal to

.5 Flow Meters

As used in the AABC National Standards, 1982, flow meters and gages are defined as follows:

Flow Meter
A device used for measuring fluid flow quantities. A flow meter generally requires a gage for indication.

Gage
An instrument used to indicate differential pressure.

A. In general, flow meters should be provided at all terminals.
B. Flow meters should be so located as to

14.6
Chapter 14—Balancing Devices

allow the flows through the mains to be balanced.

C. Locate hydronic meters so that there is no disturbance from adjacent fittings or equipment.

D. All hydronic meters must be provided with valved test connections.

E. See Chapter 9 of the AABC National Standards, 1982 for sizing of flow meters for accuracy.

.6 Coils (Fig. 14.4, 14.5)

Balancing devices must be provided for all coils as shown in Figs. 14.4 and 14.5.

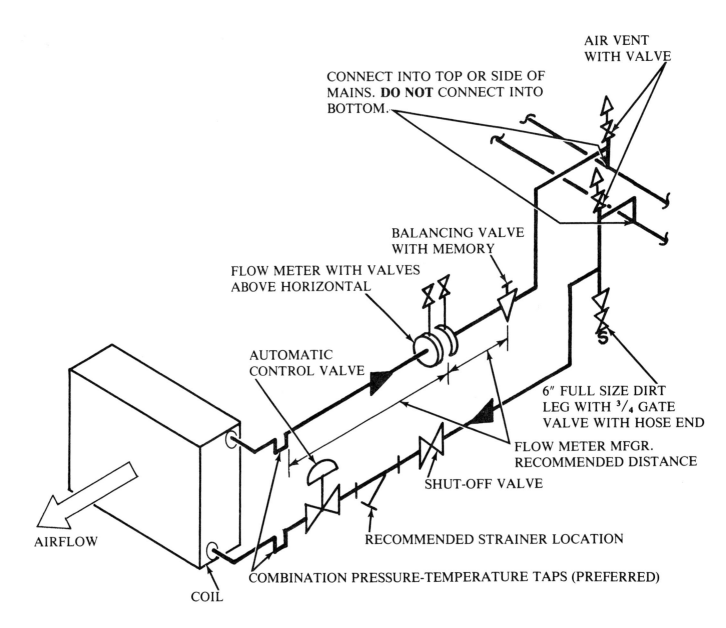

Fig. 14.4: Composite coil diagram with two-way automatic control valve

SECTION III—DESIGN STANDARDS FOR TOTAL SYSTEM BALANCE (TSB)

14.7

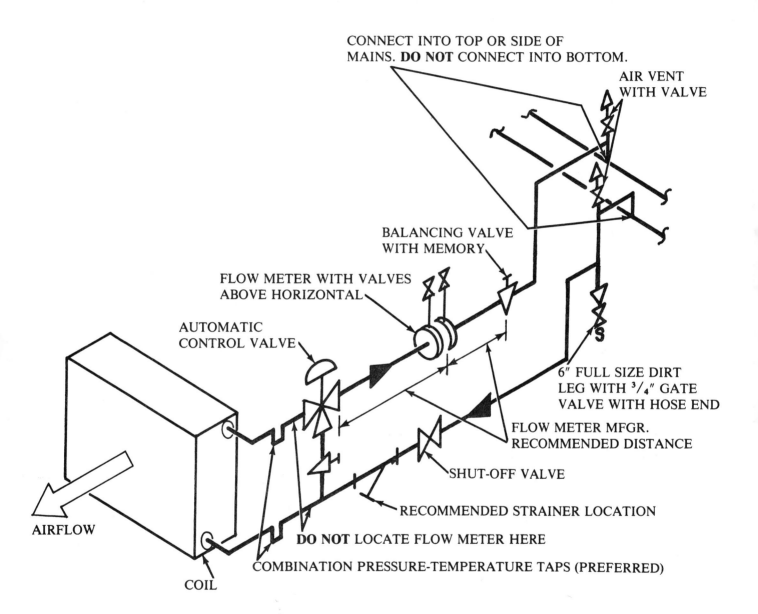

Fig. 14.5: Composite coil diagram with three-way automatic control valve

SECTION III—DESIGN STANDARDS FOR TOTAL SYSTEM BALANCE (TSB)

CHAPTER 15

SYSTEMS

15.1 OVERVIEW

Total System Balance for an air conditioning system is one part of a complex process that results in an efficient, economical system. **Total System Balance** must begin with the design of the system.

The intent of this chapter is to recommend those aspects of the Test and Balance procedure that should be considered in the original system design. Consideration given to **Total System Balance** at the design stage will result in a system that can be balanced more effectively and at a reasonable cost.

This approach is in accordance with the entire thrust of the AABC National Standards, 1982. In Section IV, Specifications, which follows this chapter, each of the chapters recommends specific balancing procedures that will achieve the Engineer's objectives for the system.

To prepare a detailed specification for **Total System Balance** without providing the means for achieving the desired objectives can only result in a less effective system or additional work to correct the deficiencies at considerable expense.

15.2 CONSIDERATIONS FOR ALL AIRFLOW SYSTEMS

.1 Items for the Specifications

A. Insulation shall not cover any nameplates or calibration data.
B. Equipment nameplate shall not be painted over or removed.
C. For coordination between the Temperature Control Contractor and the AABC Test and Balance Agency see Chapter 24.

.2 Items for the Design Drawings

A. In calculating the friction losses in an air distribution system, the Engineer has predicted the static pressure at various points. This data should be included in the design drawings. This information is important to the AABC Test and Balance Agency as a continuous check of the operation of the system, and as a means of identifying developing problems. For the Engineer, a comparison of the predicted and of the actual measured values provides an excellent check of the original design, and a means of detecting problems in the construction of the system.
B. The design drawings should include a reasonable estimate of the airflow losses due to leakage. In modern construction of duct systems, leakage is an important factor.

In systems that utilize masonry shafts and mechanical equipment spaces as return air and exhaust airflow paths, consideration of leakage as a design problem is very important.

C. Fans installed in parallel should have tight fitting, low leakage, automatic dampers installed on the discharge side of the fan. This is of utmost importance where one or more fans are installed as standby units, or if fans have been installed to accommodate future expansion. Without dampers, airflow can bypass around an inoperative fan and can cause free-wheeling in reverse rotation creating a hazard to the fan.
D. Accessibility to control devices demands adequate access doors:
 • Near all dampers.

- At all air terminal equipment such as VAV boxes and mixing boxes.
- On both sides of coils, for ease in inspection of the coil.
- At each zone of multizone units for inspection and adjustment of dampers for tight shut-off.

It is usually not enough to specify in general terms that access doors must be installed at all control devices. The design drawings should have the access openings shown in the required locations.

E. It is also important to coordinate the mechanical design requirements with the architectural design. A 12" × 12" opening in a fixed inaccessible suspended ceiling will not allow a service person to get into a ceiling space to service or adjust an air terminal unit such as a VAV unit or a constant volume mixing box.

F. Provide for sealing off floors from each other as completely as possible to prevent stack effect.

G. Calculate system resistance as accurately as possible. Include:
  1. Friction
  2. Dynamic losses
  3. System effect

15.3 AIR SYSTEMS
.1 General

A. The building structure should not be used as an air duct since building materials are porous and cause a high percentage of leakage. Use approved duct materials. If such structures must be used, use extreme care that the entire surface and connections are sealed.

B. Duct liner must be installed by a method that assures that it will stay in place. All leading edges of the liner must be protected to prevent eroding.

C. Do not apply duct liner in the discharge transition of air handling units where high velocity occurs. This restriction will result in high resistance and can cause erosion of the liner. The duct liner should not start until the location in the ductwork where normal velocity begins.

D. All supply ducts must be sealed by a sealant. Tape is not acceptable.

E. Always make calculations of the heat transmission through the duct work of the supply air system to determine where insulation is needed. This is especially critical where static regain design is used because of the increased area of the duct work per unit of airflow near the end of the system. For cooling applications, any duct with an air velocity of 1500 FPM should be insulated.

F. Supply ducts in return air plenums, such as ceiling spaces, must be insulated if the discharge air temperature at the supply air terminal is to be the same as the supply air temperature at the air handling unit.

G. A fan system can be adversely affected by connecting return or outside air to only one side of the fan inlet plenum.

H. In multizone systems, zone connections and turning vanes should be installed in such a manner as to prevent stratification of temperatures.

I. Any zone of a multizone system that utilizes long duct runs should be oversized. Runs that require more than 1.25 modules of multizone unit outlet should be oversized to two modules to minimize air distribution problems. Any shortage of outlet modules should be absorbed by the larger zones.

J. Assure that the air delivered to a conditioned space does not have to pass through a different space or zone to reach a return air grille.

K. The cooling coil capacity of any VAV system must equal the block load capacity of the fan.

L. Specify that the Temperature Control Contractor is to cooperate with the AABC Agency by assisting in setting the controls to the modes necessary for Testing and Balancing.

SECTION III—DESIGN STANDARDS FOR TOTAL SYSTEM BALANCE (TSB)

M. Assure that discharge air from roof top units such as cooling towers and condensing units cannot be introduced into the outside air intakes and carried back into the building.

N. A return air fan is recommended for systems utilizing an economizer cycle. If return/exhaust air fans are not used, either exhaust or relief air fans should be used to limit building pressurization to approximately 0.05" WG.

O. Intake air louvers should have removable screens of hardware cloth of $1/2"$ openings or larger.

P. All exhaust ducts should be sealed.

.3 Duct Design

A. All ductwork shall be designed in accordance with SMACNA duct design manuals.

B. Specify that duct connections and air turns are to be made in accordance with the appropriate SMACNA Standards. All connections should be as typical as possible.

C. No double thickness turning vanes are to be used.

D. Avoid branch take-offs close to elbows.

E. Use bell-mouth taps with round pipe. With rectangular duct, use a 90° tap with a 45° entry.

F. If flex duct is used to connect diffusers, they must be secured by drawbands and screws. Duct tape is not acceptable. Flex duct must be supported level and devoid of sharp bends to avoid excessive friction losses.

G. All square-throat 90° elbows in supply, return, and exhaust ducts must have turning vanes.

H. Provide means for true mixing of outside air and return air.

I. In high rise buildings, dampers are to be provided at each floor in return air ducts.

J. All automatic dampers shall be capable of tight close-off.

K. Provide straight runs of duct wherever pitot tube traverses are expected.

L. If the air in exhaust ducts carries materials (such as lint) only radius-throat elbows without turning vanes should be used.

M. On Variable Primary/Variable Secondary, Powered Induction Pressure Independent Systems (see Chapter 20), the air from the powered induction box should enter the common duct at 90°. Make provision for mixing the two airstreams. Discharge ducts with a tee fitting and a splitter damper are not satisfactory.

N. In Variable Air Volume Systems, determine if heating is necessary in the mixed air plenum in order to prevent the supply air temperature from being too low during periods of low supply air quantity.

.4 Duct Outlets

A. Do not mix low pressure and high pressure drop terminals (grilles, diffusers, light troffers) on the same run of duct. This makes balancing impossible. Design the system so that high-pressure and low-pressure drop terminals are on separate runs.

B. Air outlets should not be located over areas that preclude accessibility for balancing.

C. Do not use combination supply-return outlets. The percent of supply air that is short circuiting to return cannot be determined.

D. Return air grilles should operate at low face velocities (100-600 FPM) to minimize noise.

E. Specify that VAV terminals—and all others with factory installed volume regulators that are sensitive to air entering conditions—must be connected with straight duct for the length recommended by the manufacturer. Anything other than a straight air entry will result in erratic operation of the volume regulator.

.5 Drive Trains
A. Specify that V-belt drives shall be used with a horsepower capacity rating of not less than 1.5 times the motor rating. Submittals on blowers should include details of the drives.
B. No 3V, 5V, or 8V belts and sheaves should be allowed.
C. No solid belt guards shall be used. Use expanded metal for the screen of the guards. It is strongly recommended that the screen be hinged or otherwise arranged to allow quick access to the drive components. Since access will be necessary frequently, this will reduce the cost of both balancing and maintenance.
D. Belt guards should be floor mounted with no connection to the equipment.
E. Specify that a 2" diameter hole be provided in the belt guard screen in line with the fan shaft to allow tachometer readings without removing the guard.

15.4 HYDRONIC SYSTEMS (Fig. 15.1)
A. Provide for accessibility to valves and flow meters.
B. Specify that piping insulation shall not cover the removal section of strainers.
C. Specify that piping must be clean when installed.
D. Specify that the system must be thoroughly flushed before **Total System Balance** is started.
E. Specify that temporary cross-over piping with a strainer shall be installed between the supply and return lines in order to capture debris during flushing. The location of the cross-over shall be at the most probable point of debris accumulation in the supply runs.
F. Provide for dirt pockets in piping before all coils.
G. Specify the proper sleeves for all strainers for the fluid being handled.
H. Do not use three-way valves on reheat coils. When heating is not required, heat is still carried from the valve (which is constantly hot) by conduction. This will raise the temperature of the supply air.
I. Specify that all pumps have casings with gage tappings for both suction and discharge pressures.
J. Never allow pumps to operate under negative pressure without special consideration because of the danger of entrapped air.
K. Where primary-secondary pumping is used, a common pipe must be installed in either the primary or secondary circuit to eliminate pressure interaction between all pumps in the system. (Fig. 15.1)
L. Whenever parallel pumps are used, check valves must be installed in the discharge piping of each pump.

SECTION III—DESIGN STANDARDS FOR TOTAL SYSTEM BALANCE (TSB)

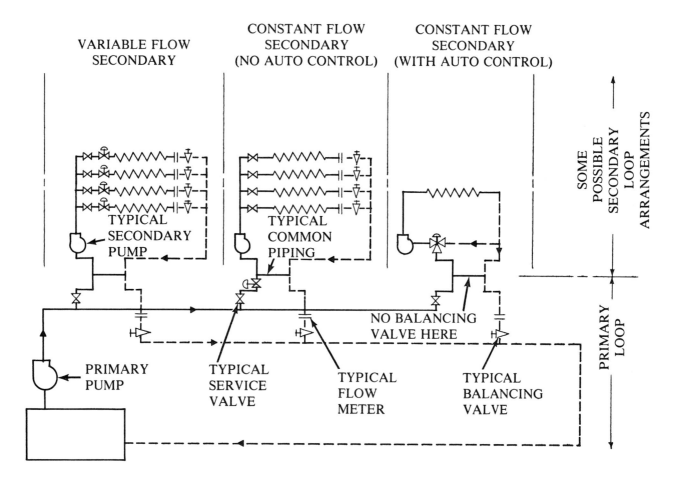

NOTES:
1. NO AUTOMATIC CONTROL VALVES IN BRANCH PIPING PROVIDE CONSTANT FLOW PRIMARY LOOP.
2. AUTOMATIC CONTROL VALVES IN BRANCH PIPING PROVIDE VARIABLE FLOW PRIMARY LOOP.

Fig. 15.1: Primary/secondary piping connections

CHAPTER 16

AABC GENERAL SPECIFICATIONS

16.1 OVERVIEW

This chapter contains the general specifications for **Total System Balance.** Use this chapter as the general specifications. Then add material from Chapters 17 through 24 for specific systems that apply to the project.

Comprehensive and detailed specifications that apply to the project are essential for thorough, professional **Total System Balance.** This can be accomplished by proper application of the specifications in these chapters.

16.2 SCOPE OF WORK

.1 **Total System Balance** shall be performed by an agency certified by the Associated Air Balance Council (AABC) and approved by the Owner's Representative. All work done by this agency shall be by qualified Technicians under the direct supervision of an AABC Certified Test and Balance Engineer.

.2 **Total System Balance** shall be performed in accordance with the latest edition of the AABC National Standards, 1982 for **Total System Balance**, and in accordance with the scope of work specified in the contract documents.

.3 **Total System Balance** shall not begin until systems are complete.

.4 Upon the completion of the work, the Test and Balance Agency shall submit _____ copies of the complete Test and Balance Report.

.5 One agency shall be responsible for all phases of **Total System Balance.**

.6 The responsibility for performing **Total System Balance,** as defined by ASHRAE, is "The overall concept requires that one source be responsible for the complete testing, adjusting, and balancing of all systems." (ASHRAE 1980 SYSTEMS, pg. 40.1).

.7 The Test and Balance Agency shall permanently mark the settings of all valves, dampers, and other adjustment devices in a manner that will allow the settings to be restored. If a balancing device is provided with a memory stop, it shall be set and locked.

16.3 SUBMITTALS

.1 The name of the Test and Balance Agency, plus the name and registration number of the Certified Test and Balance Engineer, shall be submitted to the Owner's Representative for approval within 30 days after the award of the project contract.

.2 The selected Test and Balance Agency shall submit to the Owner's Representative:
A. Detailed procedures
B. Agenda
C. Report Forms
D. AABC National Project Performance Guaranty

An approved copy of each of the above must be returned to the Test and Balance Agency before **Total System Balance** is begun.

.3 If a complete submittal in accordance with paragraph 16.3.2 is not received within the specified time, the Owner's Representative reserves the right to select the AABC Test and Balance Agency.

16.4 WORK OF OTHER TRADES

.1 The Contractor shall provide the Test and Balance Agency with one set of the following documents:
A. Within 30 days after approved selection

of the Test and Balance Agency:
1. Contract drawings
2. Applicable specifications
3. Addenda
B. As issued:
1. Change orders
C. Within 30 days after approval of the below items:
1. Approved shop drawings
2. Approved equipment manufacturer's submittal data
3. Approved temperature control drawings

.2 The Test and Balance Agency shall be provided with:
A. Reasonable time, as determined by the Test and Balance Agency, to complete Test and Balance prior to the specified completion date
B. Completely operable systems
C. The right to adjust the systems
D. Access to system components
E. Master keys if the building is occupied
F. Secure storage space for tools and instruments

.3 The Contractor is responsible for start-up and operation of systems during **Total System Balance**. Start-up shall include the following:
A. All equipment operable in safe and normal condition
B. Temperature control systems installed complete and operable
C. Proper thermal overload protection in place for electrical equipment
D. Air systems
1. Final filters clean and in place. If conditions warrant, the Contractor shall install temporary media in addition to the final filters.
2. Duct systems clean of debris
3. Correct fan rotation
4. Fire and volume dampers in place and open
5. Coil fins cleaned and combed
6. Access doors closed and duct end caps in place
7. All outlets installed and connected
8. Duct system leakage shall not exceed the rate specified
E. Hydronic systems
1. Flushed, filled, and vented
2. Correct pump rotation
3. Proper strainer baskets clean and in place
4. Temporary start-up strainer baskets removed
5. Service and balance valves open

.4 If it is determined by the Test and Balance Agency that drive changes are required, the purchaser of the equipment must obtain and install all necessary new components.

16.5 GENERAL REQUIREMENTS

.1 The work required shall be as hereinafter specified.

.2 Pre-construction plan checks and (<u>number</u>) mechanical construction reviews shall be provided by the Test and Balance Agency.

16.6 GENERAL BALANCING PROCEDURES

.1 The Test and Balance Agency shall cooperate with the Owner's Representative and all Contractors to perform the work in such a manner as to meet the job schedule, providing that sufficient lead time for **Test and Balance** has been allowed.

.2 The Test and Balance Agency shall leave all system components in proper working order, such as:
A. Replace belt guards
B. Close access doors
C. Close doors to electrical switch boxes
D. Restore thermostats to specified settings.

.3 All recorded data shall represent a true, actually measured, or observed condition.

.4 Any abnormal conditions in the mechanical systems or conditions which prevent **Total System Balance,** as observed by the Test and Balance Agency, shall be reported as quickly as possible to the individual responsible.

SECTION IV—SPECIFICATIONS

.5 If, for design reasons, a system cannot be properly balanced, it shall be reported by the Test and Balance Agency as soon as observed.

.6 Should additional balancing devices be required, the Test and Balance Agency shall bring it to the attention of the individual responsible.

Single sheets of specifications suitable for photocopying are available from AABC National Headquarters, upon request.

CHAPTER 17

SUPPLY AIR SYSTEMS—GENERAL

17.1 OVERVIEW

The specifications in this chapter are for the **Total System Balance** for all supply air systems. The areas covered are:

- **Preparation for Total System Balance**
- Information and procedures related to:
  - Outlets
  - Supply outlet tolerances
  - Fans
  - Coils
  - Other devices
  - Temperature control dampers
  - Return air ratios
  - Mixed air control

The requirements for tolerance from room to room are very rigid, as specified in this chapter. However, the tolerance between outlets **within the same space** are less rigid without decreasing comfort (Fig. 17.1).

17.2 PREPARATION FOR TOTAL SYSTEM BALANCE

.1 **Total System Balance** shall not begin until the AABC Test and Balance Agency has verified that start-up procedures have been performed as specified in Chapter 16 of the AABC National Standards, 1982.

.2 The AABC Test and Balance Agency shall measure the amperes of all fan motors before **Total System Balance** is started and shall take proper steps to correct and report any overloads.

.3 The AABC Test and Balance Agency shall not continue **Total System Balance** if any conditions are observed that are hazardous to the air system. This shall be reported before proceeding further.

.4 The AABC Test and Balance Agency shall verify all outlets for compliance with design requirements and shall report any variations before starting **Total System Balance**.

17.3 SUPPLY FANS

.1 The AABC Test and Balance Agency shall set the fan RPM to provide design total CFM within acceptable limits as indicated in the AABC National Standards, 1982, and/or required static pressure to operate the system.

.2 Fan speed shall not exceed the maximum allowable RPM as established by the fan manufacturer.

.3 The final setting of fan RPM shall not result in overloading the fan motor in any mode of operation. Dampers shall be modulated, and the amperes of the supply fan motor shall be measured to ensure that no motor overload can occur. The amperes shall be measured in the full cooling, heating, and economizer modes to determine the maximum brake horsepower.

.4 After **Total System Balancing**, the following values shall be recorded:
  A. Fan RPM
  B. Motor voltage and amperes
  C. Entering static pressure
  D. Leaving static pressure

.5 Final RPM of the fan shall be set to supply the required CFM with filters artificially restricted to simulate 50% loading. The AABC Test and Balance Agency shall verify that the fan motor will not be overloaded when the system is operating with unrestricted, clean filters in place.

.6 When applicable, final fan settings shall be based on rated wet cooling coil resistance.

.7 Final RPM of the supply fan in systems having mixed air dampers shall be set to provide required CFM with the system in a logical non-modulating mode; for example, minimum outside air.

.8 When job conditions permit, static pressure shall be measured as follows:

Chapter 17—Supply Air Systems—General

A. Static pressure leaving the fan shall be taken as far downstream from the fan as is practical, but shall be upstream of any restrictions in the duct (such as duct turns).

B. No reading shall be taken directly at the fan outlet or through the flexible connection.

C. Static pressure entering a single inlet fan shall be measured in the inlet duct upstream of any flexible connection and downstream of any duct restrictions.

D. Static pressure entering a double inlet fan shall be measured through the wall of the plenum which houses the fan.

In all cases, the readings shall be taken so as to represent as true a value as possible. True value is actual measured static pressure.

17.4 OUTLETS (Fig. 17.1)

.1 All quantities shall be measured according to the AABC National Standards, 1982.

.2 The systems shall be balanced so that the total supply air quantity to **each space** shall be within −5% to +10% of the design amount.

.3 **Each outlet** within the same space as related to 17.4.2 above shall be adjusted to design quantities in accordance with Fig. 17.1. All final quantities shall be obtained without generating noise.

.4 The pattern for all adjustable outlets shall be adjusted for proper distribution without drafts.

.5 If, during **Total System Balance**, the Test and Balance Agency detects any outlet conditions that will not allow proper balancing to be performed, the facts shall be reported immediately.

17.5 FILTERS

Under final balanced conditions, the AABC Test and Balance Agency shall measure and record static pressure entering and leaving each filter bank.

17.6 COILS AND OTHER DEVICES

.1 Under final balanced conditions, the AABC Test and Balance Agency shall measure and record static pressures entering and leaving each coil bank.

.2 Under final balanced conditions, the AABC Test and Balance Agency shall measure and record static pressures entering and leaving other devices not normally found in a system (such as, but not limited to, sound traps, heat recovery equipment, and air washers).

17.7 TEMPERATURE CONTROL DAMPERS (AUTOMATIC)

.1 All temperature control dampers shall be verified by the AABC Test and Balance Agency for proper shut-off when driven closed by the controller. Dampers shall also be verified to be in the same position as indicated by the controller. Required corrections will be by others.

17.8 MIXED AIR CONTROL

.1 Manual balancing dampers in return, outside, and/or relief air connections shall be restricted as necessary so the system supplies and returns essentially the same CFM in any mode of modulation.

.2 The AABC Test and Balance Agency shall observe or test mixed air plenums for possible stratification. If freeze-up or other serious problems are likely, the condition shall be reported at once.

.3 The AABC Test and Balance Agency shall observe the start-up of medium and high pressure systems to check that no dangerous conditions exist. If dangerously low pressure in the fan inlet plenum, or dangerously high pressure in the fan discharge plenum are observed, they shall be reported or corrected at once.

4. The AABC Test and Balance Agency shall set the minimum outside air quantity to the required value. If this airflow quantity cannot be properly measured, the Temperature Method as specified in the AABC National Standards, 1982 shall be used.

SECTION IV—SPECIFICATIONS

	Number of Outlets in the Space		
Classification of Space	1	2	3 or more
General	-5% +10%	±10%	±15%
Warehouse or Industrial	-5% +10%	±15%	±15%
Operating Room or Other Special Environmental Rooms	+5%	±5%	±10%

Note: The values in this figure are applicable only to each outlet located within the same space. The requirements for tolerance from room to room are very rigid, as specified in this chapter. However, the tolerance between outlets **within the same space** is less rigid without decreasing comfort.

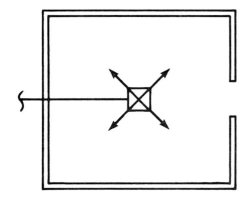

ONE OUTLET IN THE SPACE

OUTLET SUPPLY AIR QUANTITY: WITHIN -5% TO +10% OF DESIGN.

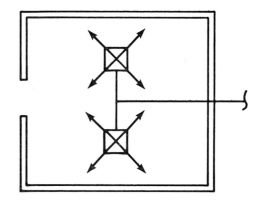

TWO OUTLETS IN THE SPACE

TOTAL SUPPLY AIR QUANTITY: WITHIN -5% TO +10% OF DESIGN. SUPPLY AIR QUANTITY OF EACH OUTLET WITHIN ±10% OF DESIGN.

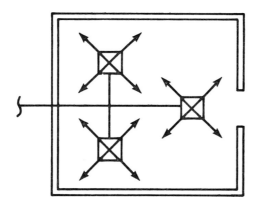

THREE OR MORE OUTLETS IN THE SPACE

TOTAL SUPPLY AIR QUANTITY: WITHIN -5% TO +10% OF DESIGN. SUPPLY AIR QUANTITY OF EACH OUTLET: WITHIN ±15% OF DESIGN.

Fig. 17.1: % Tolerance between supply air outlets within a space

CHAPTER 18

LOW PRESSURE AIR SYSTEMS

18.1 OVERVIEW

This chapter specifies the Standards for the adjustment and balancing of Low Pressure Air Systems. The Standards are categorized into:
- Single zone
- Multizone
- Reheat or recool systems

18.2 SINGLE ZONE SYSTEM (Fig. 18.1)

.1 At completion of balancing, at least one outlet damper shall be fully open on every branch duct.

.2 At completion of balancing, at least one branch duct balancing damper shall be fully open.

.3 Airflow quantity of the fan shall be determined by pitot tube traverse unless impractical to do so. Traverses shall be taken as close to the fan as allowed by the AABC National Standards, 1982. When the quantity cannot be obtained by pitot tube traverse, the summation of the outlet quantities shall be used as the total CFM of the fan. Information shall be so noted on the data sheet.

.4 Static pressure shall be measured at the points labeled "P" in Fig. 18.1.

18.3 MULTIZONE SYSTEM (Fig. 18.2)

.1 At completion of balancing, at least one outlet damper shall be fully open on every branch duct.

.2 At completion of balancing, at least one branch duct balancing damper shall be fully open on every zone duct.

.3 At completion of balancing, at least one zone balancing damper shall be fully open.

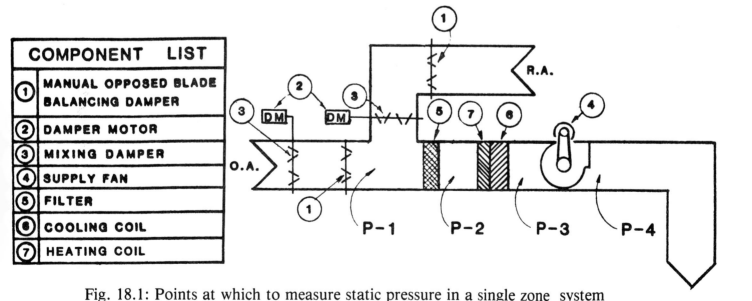

Fig. 18.1: Points at which to measure static pressure in a single zone system

Chapter 18—Low Pressure Air Systems

.4 Final measured data shall be taken with the zone dampers in a non-modulating mode as applicable to the system.

.5 The zone temperature control dampers shall be verified by the Test and Balance Agency for proper shut-off of both hot and cold decks. It shall also be verified that all zone mixing dampers are controlled by the proper space thermostat.

.6 Zone dampers shall be modulated, and the amperes required by the supply fan motor shall be measured to ensure that no motor overload can occur. Check in full heating, full cooling, and economizer modes to determine where maximum brake horsepower is required.

.7 Total airflow quantity for each zone shall be determined by pitot tube traverse unless impractical to do so. Traverses shall be taken as close to the unit as allowed by the AABC National Standards, 1982. Where the quantity cannot be obtained by pitot tube traverse, the summation of the outlet quantities shall be used as the total CFM of the zones. Information shall be so noted on the data sheet.

.8 Static pressure shall be measured at the points labeled "P" in Fig. 18.2.

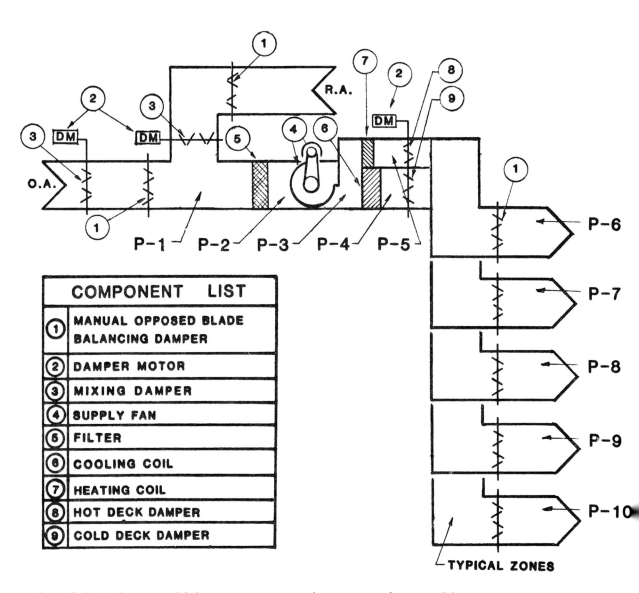

Fig. 18.2: Points at which to measure static pressure in a multizone system

18.4 REHEAT OR RECOOL SYSTEM (Fig. 18.3)

.1 At completion of balancing, at least one outlet damper shall be fully open on every branch duct.

.2 At completion of balancing, at least one branch duct balancing damper shall be fully open.

.3 Total airflow quantity at each reheat zone shall be determined by pitot tube traverse unless impractical to do so. Traverses shall be taken as close to the reheat coils as permitted by the AABC National Standards, 1982. Where the quantity cannot be obtained by pitot tube traverse, the summation of the outlet quantities shall be used as the total CFM of the zones. This information shall be so noted on the data sheet.

.4 Static pressure shall be measured at the points labeled "P" in Fig. 18.3.

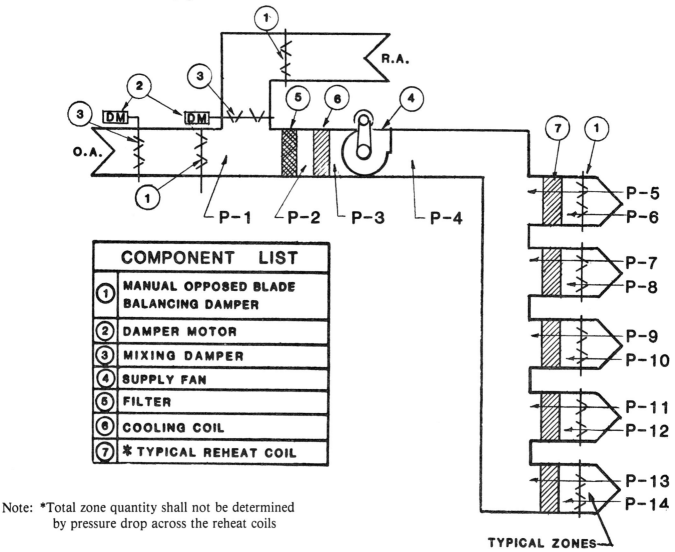

Note: *Total zone quantity shall not be determined by pressure drop across the reheat coils

Fig. 18.3: Points at which to measure static pressure in a reheat system

CHAPTER 19

MEDIUM AND HIGH PRESSURE AIR SYSTEMS

19.1 OVERVIEW

This chapter covers the specifications for the following systems:
- Induction
- Constant Volume, Single Duct
- Constant Volume, Dual Duct, with Single-motor Mixing Boxes
- Constant Volume, Dual Duct, with Two-motor Mixing Boxes

These specifications include an acceptable alternate method for determining total airflow quantities if pitot tube traverses are not practical.

The specifications in this chapter clearly make the AABC Test and Balance Agency responsible for adjustment of volume regulators of all terminals. The responsibility for repair or replacement is that of the manufacturer.

19.2 INDUCTION SYSTEM (Fig. 19.1, 19.2)

.1 Primary airflow to each terminal unit shall be determined by measuring the plenum pressures and comparing these pressures with the manufacturer's rated value.

.2 At the completion of balancing, the primary air damper of at least one induction unit (Fig. 19.1) on each branch duct shall be fully open.

.3 At completion of balancing, at least one branch balancing damper shall be fully open.

.4 Total airflow quantity shall be determined by pitot tube traverse unless impractical to do so. Where the quantity cannot be determined by a pitot tube traverse, the summation of the primary air quantities at all induction units shall be used as the total CFM of the fan. Information to this effect shall be noted in the data sheet.

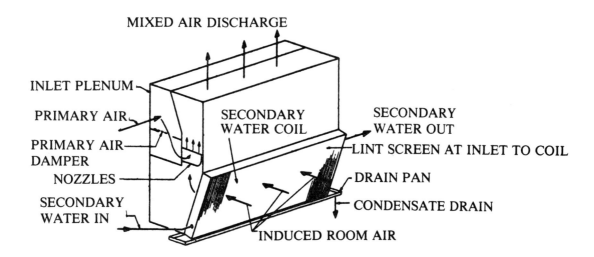

Fig. 19.1: Air-water induction unit

Chapter 19—Medium and High Pressure Air Systems

.5 Plenum static pressure of each induction unit measured under the final balanced condition shall be recorded in the report.

.6 Static pressure shall be measured at the points labeled "P" in Fig. 19.2.

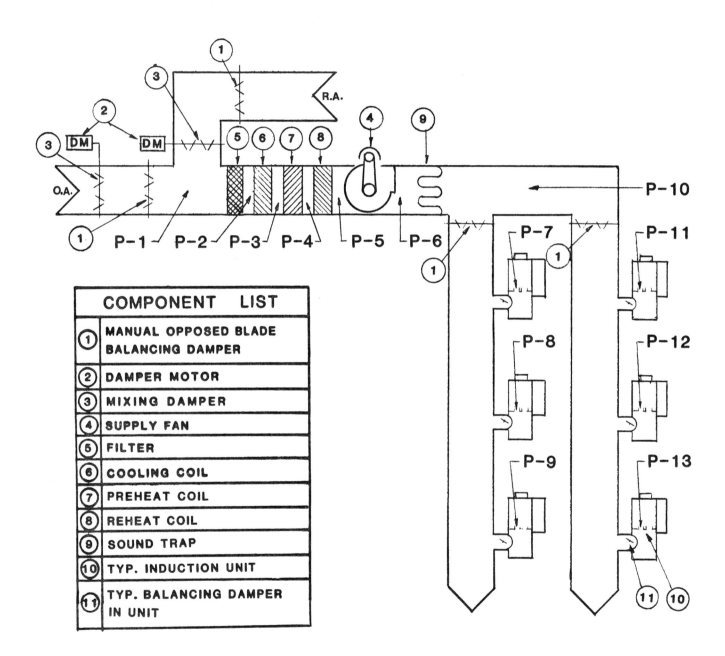

Note: P-7 through P-9 and P-11 through P-13 are typical nozzle pressures

Fig. 19.2: Points at which to measure static pressure in an induction system

19.3 CONSTANT VOLUME SINGLE DUCT SYSTEM (Fig. 19.3)

.1 The low pressure side of the system shall be balanced as specified in Chapter 18 of the AABC National Standards, 1982.

.2 As a preliminary step, the supply fan total capacity shall be determined before adjusting the constant volume devices.

    A. If the fan is capable of supplying the necessary quantity to meet the total needs of all the terminals, then each terminal must be adjusted in proportion to the available CFM. If the total fan capacity is less than 95%, the individual responsible shall be notified before proceeding with the work.

    B. If the supply fan has an adequate capacity, the constant volume regulator of each terminal shall be adjusted for the required quantity by the AABC Test and Balance Agency.

.3 **If boxes do not operate properly and repairs are required, balancing shall be suspended until corrective action is taken.**

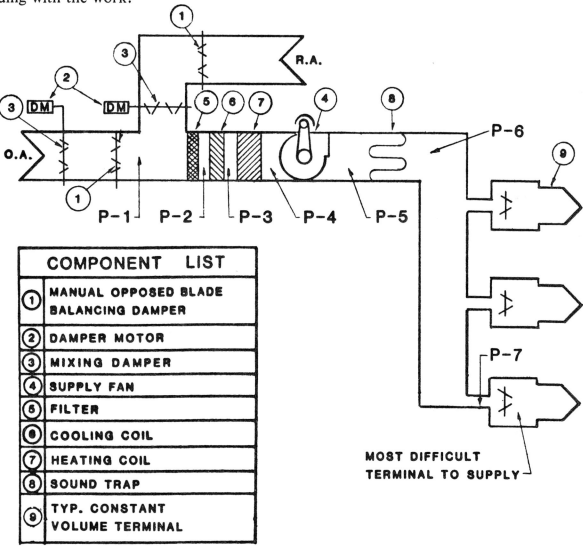

Fig. 19.3: Points at which to measure static pressure in a constant volume single duct system

.4 After all terminals are adjusted, the final capacity of the supply fan shall be set. The setting shall be such that the inlet static pressure to the terminal that was most difficult to supply will be adequate (but not in excess of what is required) to overcome the resistance of the low pressure duct work and to assure that the regulator is in control.

.5 Final total airflow quantity shall be determined by pitot tube traverse unless impractical to do so. Where the quantity cannot be determined by a pitot tube traverse, the summation of all low pressure terminals shall be used as the total CFM of the fan. Information to this effect shall be noted in AABC Report Forms.

.6 Static pressure in the duct to the most difficult-to-supply terminal shall be recorded in the Test and Balance Report.

.7 Static pressure shall be measured at the points labeled "P" in Fig. 19.3.

19.4 CONSTANT VOLUME DUAL DUCT SYSTEM WITH SINGLE-MOTOR MIXING BOXES (Fig. 19.4; 19.5)

.1 The low pressure side of the system shall be balanced as specified in Chapter 18 of the AABC National Standards, 1982.

.2 As a preliminary step, the supply fan total capacity shall be determined before adjusting the constant volume regulators.
   A. Determine the diversity factor if applicable.
   B. Set the thermostats of the boxes as required by the diversity.
   C. If in the judgment of the AABC Test and Balance Agency, the total fan capacity is less than 95%, or the static pressure is inadequate, the individual responsible shall be notified before proceeding with the work.

.3 If the supply fan has an adequate capacity, each mixing box shall be set by the AABC Test and Balance Agency to acceptable tolerance of design CFM, provided there is sufficient static pressure in the supply duct. If boxes do not operate properly and repairs are required, balancing shall be suspended until corrective action is taken.

.4 After all mixing boxes are adjusted, the thermostats shall be set so that the proper diversity is achieved. The final capacity of the supply fan shall be adjusted so that the inlet static pressure to the box that is most difficult to supply is proper to overcome the downstream resistance. This also will assure that the constant volume regulator is in control.

.5 The final total system airflow shall be determined by pitot tube traverses of the main hot and cold ducts, unless impractical to do so. Where the quantity cannot be determined by pitot tube traverse, the summation of all low pressure terminals shall be used as the total CFM of the fan. Information to this effect shall be noted in the AABC Report Form.

.6 Static pressure in the branch duct to the box which is most difficult to supply shall be recorded in the Test and Balance Report.

.7 Final measured data shall be obtained with the system in the same mode as when the fan capacity was adjusted.

.8 The zone temperature control dampers shall be verified by the Test and Balance Agency for proper shut-off of both hot and cold decks. It shall also be verified that all zone mixing dampers are controlled by the proper space thermostat.

.9 Static pressure shall be measured at the points labeled "P" in Fig. 19.5.

19.5 CONSTANT VOLUME DUAL DUCT SYSTEM WITH TWO-MOTOR MIXING BOXES
(Figs. 19.6; 19.7)

.1 The low pressure side of the system shall be balanced as specified in Chapter 18 of the AABC National Standards, 1982.

.2 The CFM of each mixing box shall be set to design quantity by the AABC Test and Balance Agency by adjusting the constant volume regulator with the room thermostat in a full cooling mode.

.3 After the regulator is adjusted for each mixing box, the thermostat shall be placed in a full heating mode and the operation verified.

.4 After all mixing boxes are adjusted, the thermostats shall be set so that the proper diversity is achieved. The capacity of the supply fan shall be adjusted so that there is sufficient pressure in the branch duct to the mixing box which is most difficult to supply. The constant volume regulator shall be in control with the cold damper nearly full open.

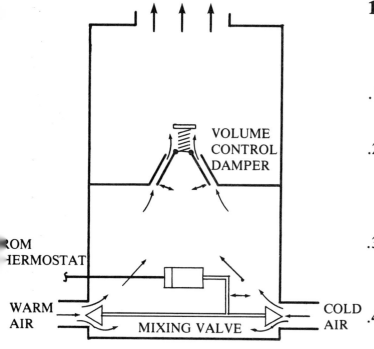

Fig. 19.4: Typical single motor mixing box

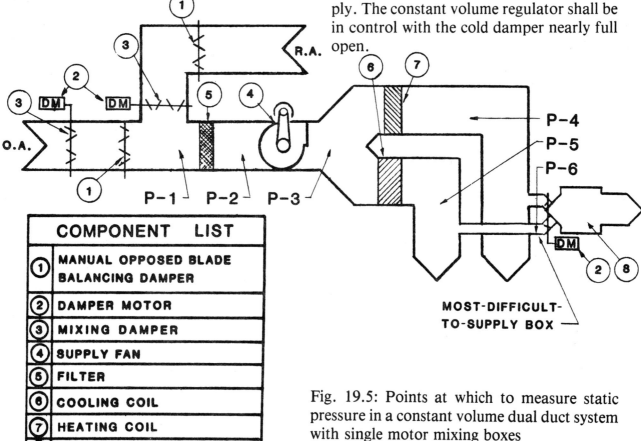

Fig. 19.5: Points at which to measure static pressure in a constant volume dual duct system with single motor mixing boxes

.5 Final total system airflow shall be determined by pitot tube traverses of the main hot and cold ducts unless impractical to do so. Where the quantity cannot be determined by pitot tube traverse, the summation of all low pressure terminals shall be used as the total CFM of the fan. Information to this effect shall be noted in the AABC Report Forms.

.6 Final measured data shall be obtained with the system in the same mode as when the fan capacity was adjusted.

.7 The zone temperature control dampers shall be verified by the Test and Balance Agency for proper shut-off of both hot and cold decks. It shall also be verified that all zone mixing dampers are controlled by the proper space thermostat.

.8 Static pressure shall be measured at the points labeled "P" in Fig. 19.7.

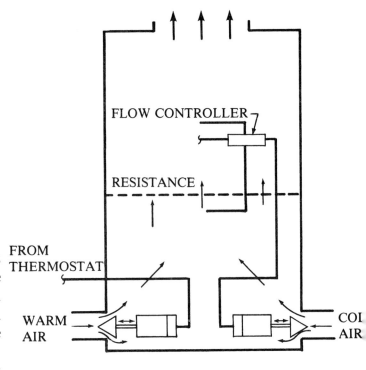

Fig. 19.6: Typical two motor mixing box

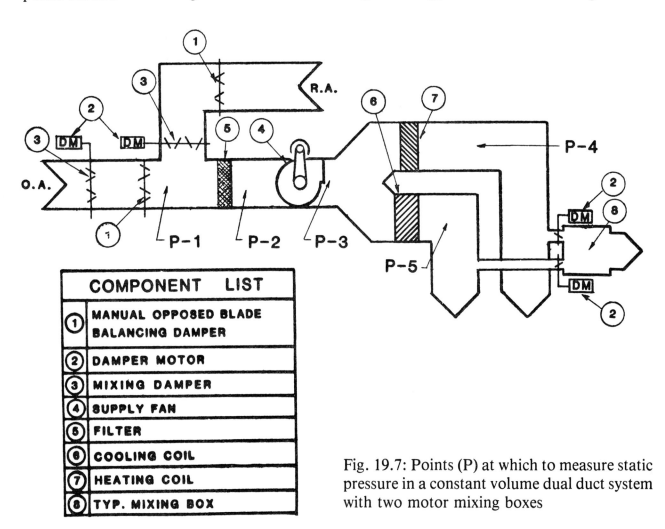

Fig. 19.7: Points (P) at which to measure static pressure in a constant volume dual duct system with two motor mixing boxes

CHAPTER 20

VARIABLE AIR VOLUME SYSTEMS

20.1 OVERVIEW (Fig. 20.1)

VAV systems are so complex and so varied in design that one set of specifications cannot be written to cover all situations. This chapter contains the specifications for Variable Air Volume Systems (VAV). To provide maximum applicability, the specifications in this chapter are divided as follows:

- 20.2 Secondary System Balancing
- 20.3 Single Duct, Variable Primary/Variable Secondary, Pressure Dependent Systems
- 20.4 Single Duct Variable Primary/Variable Secondary, Pressure Independent Systems
- 20.5 Single Fan, Dual Duct, Variable Primary/Variable Secondary, Pressure Independent Systems
- 20.6 Variable Primary/Constant Secondary, Pressure Independent, Powered Terminal Systems
- 20.7 Variable Primary/Constant Secondary, Pressure Dependent, Powered Terminal Systems
- 20.8 Variable Primary/Induction Secondary, Pressure Independent Systems
- 20.9 Variable Primary/Variable Secondary, System Powered Bladder Terminal Systems
- 20.10 Constant Primary/Variable Secondary Systems
- 20.11 Variable Primary/Variable Secondary, Parallel Powered Induction, Pressure Independent Systems
- 20.12 Supply Fan Controls for Variable Primary Systems
- 20.13 Return/Exhaust Fan For Variable Primary Systems
- 20.14 Relief Fan for Variable Primary Systems

Greater understanding of VAV systems is required of the Engineer and of the Test and Balance Agency than for any other type of HVAC systems. The Engineer **must** see that proper balancing devices are provided in the correct locations. See Chapter 14 of the AABC National Standards, 1982.

A well designed VAV system will provide a substantial degree of individual temperature control throughout the system as well as economy of operation. It will operate at lower airflow and lower power consumption than a constant volume system of similar capacity. However, there is a misconception that a VAV system is self-balancing. Because of the flexibility of the operation and the diversity that allows for a reduced total airflow, it is essential that the **Total System Balance** be accurate and complete.

If a project involves a VAV system that is not covered in this chapter, the Engineer should contact a local AABC Test and Balance Agency or National Headquarters for advice on balancing devices and on **Total System Balance** specifications.

The term **Design Diversity Factor** as used in this chapter is defined in Fig. 20.1

$$\text{Design Diversity Factor} = \frac{\text{Design Fan Capacity}}{\text{Mathematical Summation of Design Quantities of all Terminals}}$$

Fig. 20.1: Design diversity factor

20.2 SECONDARY SYSTEM BALANCING

.1 Low pressure secondary systems shall be balanced as specified in Chapter 18 of the AABC National Standards, 1982.

.2 If reheat is provided in a system, the Test and Balance Agency shall verify the proper sequence of operations of all automatic control systems applicable to the reheat system. Automatic control adjustments necessary shall be by others.

20.3 SINGLE DUCT, VARIABLE PRIMARY/VARIABLE SECONDARY, PRESSURE DEPENDENT SYSTEMS (Fig. 20.2)

.1 Space thermostats shall be set in either a full heating or cooling position as required to satisfy the Design Diversity Factor of the system, if applicable. Thermostats to be set for heating or cooling shall be selected by the Test and Balance Engineer and shall simulate as nearly as practical the manner in which the system will respond to the cooling load shift of the building.

.2 With proper diversity established, adjust the supply fan capacity to provide design total CFM with its automatic volume control device fully open. Total CFM shall be determined by pitot tube traverse. If the traverse indicates excessive duct system leakage, this condition shall be recorded in the AABC Report Form.

.3 All VAV terminals shall be proportionately balanced to receive the same ratio of required quantities of primary air. This shall be done by adjusting the manual damper in the inlet duct with all terminals in a full cooling mode (maximum CFM).

.4 Where VAV boxes with minimum CFM are provided, the stop of the automatic control damper shall be set by the Test and Balance Agency so the minimum supply quantity is as specified. Both maximum and minimum quantities shall be included in the AABC Report Form.

.5 At completion of balancing, the inlet manual damper to at least one VAV terminal on each branch duct shall be fully open.

.6 At completion of balancing, at least one damper in each branch duct shall be fully open.

.7 Static pressure shall be measured at the points labeled "P" in Fig. 20.2.

SECTION IV—SPECIFICATIONS

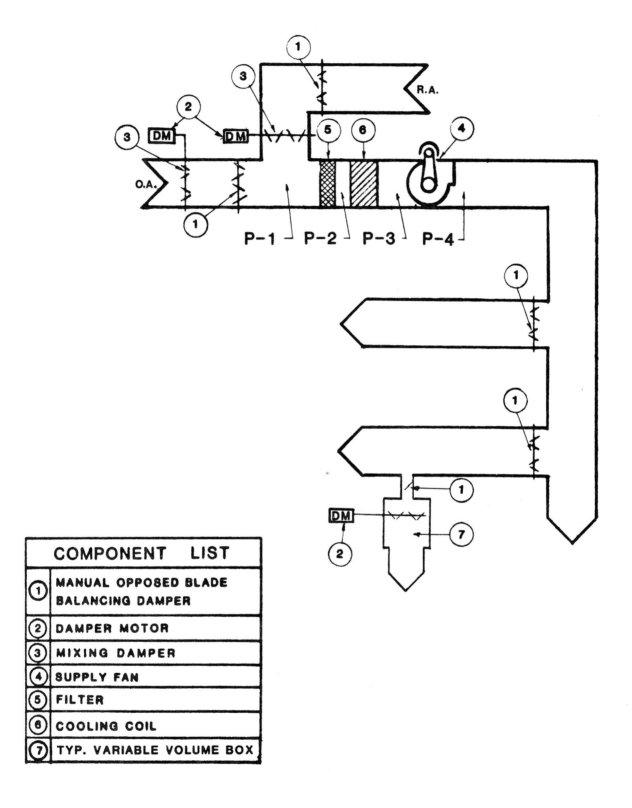

Fig. 20.2: Points at which to measure static pressure in a Single Duct, Variable Primary/Variable Secondary, Pressure Dependent VAV System

20.4 SINGLE DUCT, VARIABLE PRIMARY/VARIABLE SECONDARY, PRESSURE INDEPENDENT SYSTEMS (Fig. 20.3)

.1 The maximum volume regulator of each VAV terminal shall be set by the AABC Test and Balance Agency to supply design CFM within acceptable tolerance and with the space thermostat calling for full cooling. There must be adequate static pressure in the primary air duct to overcome the resistance of the secondary ductwork and assure that the regulator is in control.

.2 Where VAV boxes with minimum CFM are specified, the regulator shall be set to supply the required minimum quantity. Both maximum and minimum quantities shall be recorded in the AABC Report Form.

.3 If boxes do not operate properly and repairs are required, balancing shall be suspended until corrective action is taken.

.4 Set each branch damper so that the branch downstream air pressure in each branch is identical when the system is at maximum flow. The damper in the branch that is most difficult to supply shall be left open.

.5 After all VAV terminals are adjusted, space thermostats shall be set in either a full heating or cooling position as required to satisfy the Design Diversity Factor of the system, if applicable. Thermostats to be set for heating or cooling shall be selected by the Test and Balance Engineer and shall simulate as nearly as practical the manner in which the system will respond to the cooling load shift of the building.

.6 With proper diversity established, adjust the supply fan capacity with its automatic volume control device fully open to provide adequate (but not excessive) static pressure in the branch duct to the VAV terminal which is most difficult to supply.

.7 Total CFM of the supply air system shall be determined by pitot tube traverse. If the traverse indicates excessive duct system leakage, this condition shall be recorded in the AABC Report Form.

.8 If not practical to determine the CFM by pitot tube traverse, use the summation of the outlets on each terminal. System must be set in the diversity mode. Do not use the summation of **ALL** the terminals as the supply fan total maximum capacity.

.9 Static pressure shall be measured at the points labeled "P" in Fig. 20.3.

20.5
SECTION IV—SPECIFICATIONS

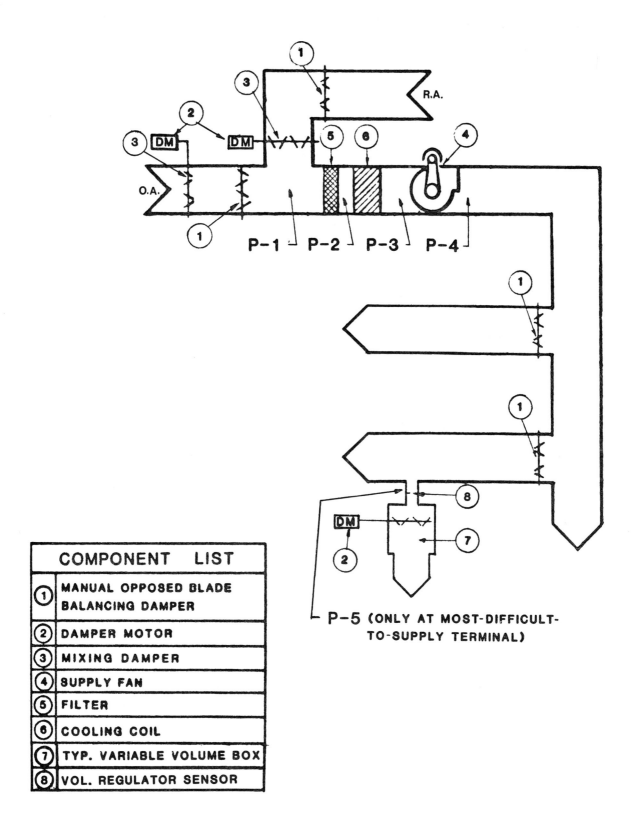

Fig. 20.3: Points at which to measure static pressure in a Single Duct, Variable Primary/Variable Secondary, Pressure Independent VAV System

20.5 SINGLE FAN, DUAL DUCT, VARIABLE PRIMARY/ VARIABLE SECONDARY, PRESSURE INDEPENDENT SYSTEMS (Fig. 20.4)

.1 This type of system utilizes a deadband control scheme which supplies a varying quantity of either heated or cooled (no mixing) air to the space through a common secondary air duct.

.2 The maximum volume regulator of both the hot and cold valves shall be set by the AABC Test and Balance Agency to supply design CFM within acceptable tolerance with the space thermostat set in full heating and cooling positions respectively. There must be adequate static pressure in the primary air ducts to overcome the resistance of the secondary ductwork and assure that the regulators are in control.

.3 If boxes do not operate properly and repairs are required, balancing shall be suspended until corrective action is taken.

.4 After all VAV terminals are adjusted, space thermostats shall be set in either full cooling or neutral (deadband with no heating or cooling) position as required to satisfy the Design Diversity Factor of the system if applicable. Thermostats to be set for cooling or neutral shall be selected by the Test and Balance Engineer and shall simulate as nearly as practical the manner in which the system will respond to the cooling load shift of the building.

.5 With proper diversity established, adjust the supply fan capacity with its automatic volume control device fully open to provide adequate (but not excessive) static pressure in the cold branch air duct to the VAV terminal which is most difficult to supply. This static pressure shall be recorded on the AABC Report Form.

6. Total CFM of the supply air system shall be determined by pitot tube traverse. If the traverse indicates excessive duct system leakage, this condition shall be recorded in the AABC Report Form.

7. If not practical to determine the CFM by pitot tube traverse, use the summation of the outlets on each terminal. System must be set in the diversity mode. Do not use the summation of **ALL** the terminals as the supply fan total maximum capacity.

.8 Static pressure shall be measured at the points labeled "P" in Fig. 20.4.

SECTION IV—SPECIFICATIONS

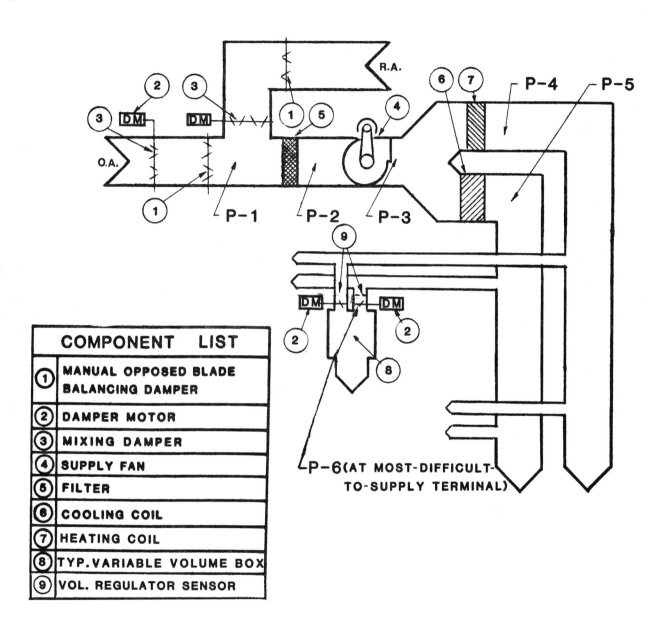

Fig. 20.4: Points at which to measure static pressure in a Single Fan, Dual Duct, Variable Primary/Variable Secondary, Pressure Independent VAV System

20.6 VARIABLE PRIMARY/ CONSTANT SECONDARY, PRESSURE INDEPENDENT, POWERED TERMINAL SYSTEMS (Fig. 20.5)

.1 This type of system utilizes a secondary fan with modulating primary air damper and barometric return air damper. The secondary fan overcomes the resistance of the secondary system and provides constant airflow to the space. The room thermostat positions the primary air damper through the high-limit volume regulator.

.2 The output of each secondary airflow shall be set at design CFM in a full return air mode if means are provided.

.3 The maximum volume regulator of each VAV terminal shall be set by the AABC Test and Balance Agency to supply design CFM within acceptable tolerance and with the space thermostat calling for full cooling. There must be adequate static pressure in the primary air duct to assure that the regulator is in control.

.4 If boxes do not operate properly and repairs are required, balancing shall be suspended until corrective action is taken.

.5 Set each branch damper so that the downstream branch air pressure in each branch is identical when the system is at maximum flow. The damper in the branch that is most difficult to supply shall be left open.

.6 After all VAV terminals are adjusted, space thermostats shall be set in either a full heating or a full cooling position as required to satisfy the Design Diversity Factor of the system, if applicable. Thermostats to be set for heating or cooling shall be selected by the Test and Balance Engineer and simulate as nearly as practical the manner in which the system will respond to the cooling load shift of the building.

.7 With proper diversity established, adjust the supply fan with its automatic volume control device fully open, to provide adequate (but not excessive) static pressure in the branch duct to the VAV terminal which is most difficult to supply.

.8 Total CFM of the supply air system shall be determined by pitot tube traverse, if practical. If the traverse indicates excessive duct system leakage, this condition shall be recorded on the AABC Report Form.

.9 If not practical to determine the CFM by pitot tube traverse, use the summation of the outlets on each terminal. System must be set in the diversity mode. Do not use the summation of **ALL** the terminals as the supply fan total maximum capacity.

.10 Static pressure shall be measured at the points labeled "P" in Fig. 20.5.

20.9
SECTION IV—SPECIFICATIONS

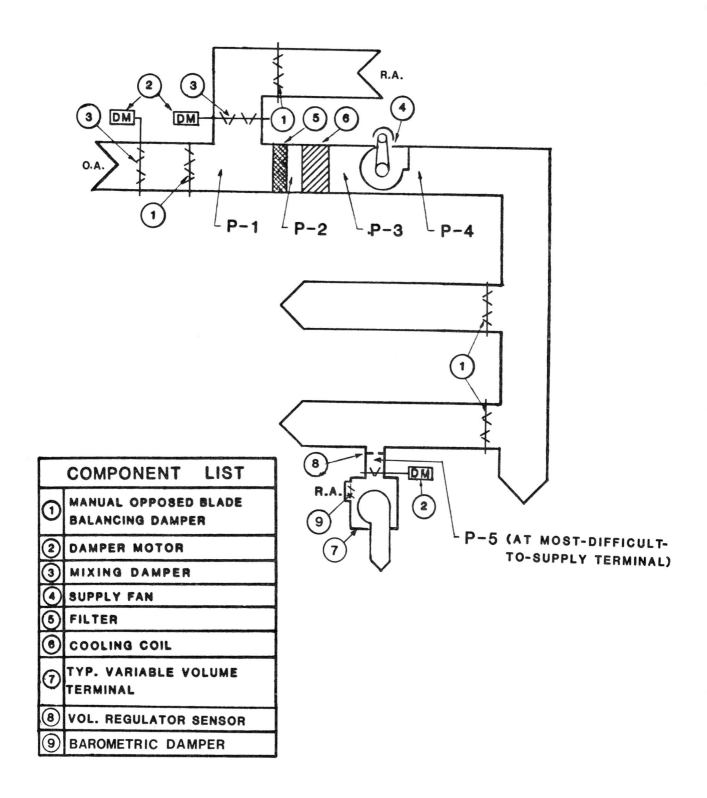

Fig. 20.5: Variable Primary, Constant Secondary, Pressure Independent, Powered Terminal VAV System

20.7 VARIABLE PRIMARY/CONSTANT SECONDARY, PRESSURE DEPENDENT, POWERED TERMINAL SYSTEMS (Fig. 20.6)

.1 This type of system utilizes a secondary fan with modulating mixing dampers (or barometric return damper) controlled by a space thermostat to overcome the resistance of the secondary system. The automatic dampers mix supply air from the primary system with return air from the ceiling space as required to satisfy the thermostat. Without a constant volume control at the VAV terminal it is difficult, if not impossible, to balance this system within the tolerances of the AABC National Standards, 1982.

.2 Set the thermostats for the number of VAV terminals to establish system diversity. With proper diversity established, balance the most difficult-to-supply VAV terminal to deliver design CFM within acceptable tolerances.

.3 The manual balancing damper of each VAV terminal shall be set by the Test and Balance Agency to supply design CFM within acceptable tolerances with the space thermostat calling for full cooling. There must be adequate static pressure at all times in the primary air duct. This assures that the negative pressures created by the secondary air fan can be offset so that no return air is induced into the secondary system.

.4 If provided, the balancing damper in the return air inlet shall be set so each terminal supplies essentially rated CFM when in the full return air mode. If no balancing or barometric damper is provided, follow the manufacturer's instructions.

.5 At completion of balancing, the manual damper in the primary supply air duct to at least one VAV terminal on each branch duct shall be fully open.

.6 At completion of balancing, at least one damper in each branch duct shall be fully open.

.7 After balancing is completed, space thermostats shall be set in either a full heating or cooling position as required to satisfy the Design Diversity Factor of the system if applicable. Thermostats to be set for heating or cooling shall be selected by the Test and Balance Engineer and shall simulate as nearly as practical the manner in which the system will respond to the cooling load shift of the building.

.8 With proper diversity established, adjust the supply fan with its automatic volume control device fully open, to provide adequate (but not excessive) static pressure in the branch duct to the VAV terminal which is most difficult to supply.

.9 Total CFM of the supply air system shall be determined by pitot tube traverse, if practical. If not, the summation of the secondary air outlets shall be used. Since this system does not have constant volume control, the results of a pitot tube traverse or the summation of the outlets may not establish fan actual CFM within acceptable tolerances as established by the AABC National Standards, 1982. This information shall be so noted on the AABC Report Form.

.10 If not practical to determine the CFM by pitot tube traverse, use the summation of the outlets on each terminal. System must be set in the diversity mode. Do not use the summation of **ALL** the terminals as the supply fan total maximum capacity.

.11 Static pressure shall be measured at the points labeled "P" in Fig. 20.6.

SECTION IV—SPECIFICATIONS

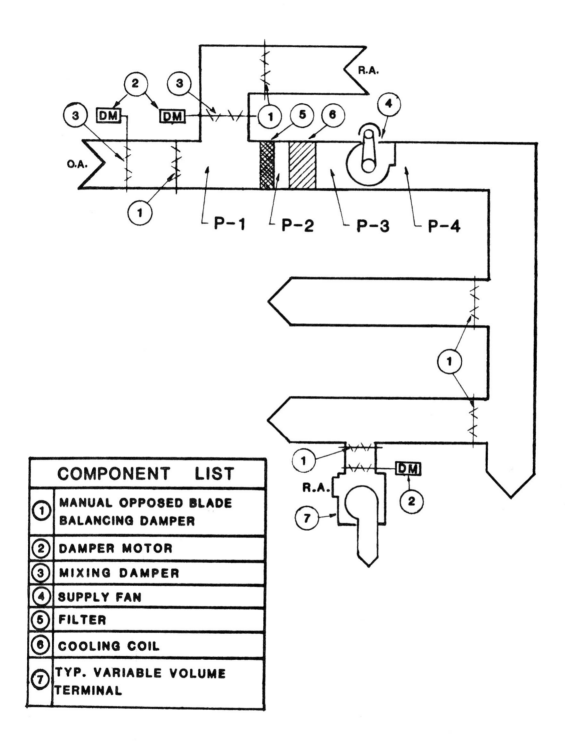

Fig. 20.6: Points at which to measure static pressure in a Variable Primary/Constant Secondary, Pressure Dependent, Powered Terminal VAV System

20.8 VARIABLE PRIMARY/ INDUCTION SECONDARY, PRESSURE INDEPENDENT SYSTEMS (Fig. 20.7)

.1 The maximum volume regulator of each VAV terminal shall be set by the AABC Test and Balance Agency to supply design CFM within acceptable tolerance and with the space thermostat calling for full cooling. There must be adequate static pressure in the primary air duct to overcome the resistance of the secondary ductwork and to assure that the box can induce return air.

.2 If boxes do not operate properly and repairs are required, balancing shall be suspended until corrective action is taken.

.3 Set each primary air branch damper so that the downstream static pressure in each branch is sufficient when the system is at maximum flow. The damper in the branch that is most difficult to supply shall be left open.

.4 With the thermostat set for full heating, the quantity of primary air shall be measured. If different from the design requirement, the damper linkage shall be adjusted as necessary.

.5 The AABC Test and Balance Report shall include the quantity of primary air supplied to each VAV terminal in both maximum primary and minimum primary modes in accordance with the manufacturer's recommendations. It is not intended that the total CFM of the box be measured at minimum primary air.

.6 After all VAV terminals are adjusted, space thermostats shall be set in either a full cooling or heating position as required to satisfy the Design Diversity Factor of the system, if applicable. Thermostats to be set for heating or cooling shall be selected by the Test and Balance Engineer and shall simulate as nearly as practical the manner in which the system will respond to the cooling load shift of the building.

.7 With proper diversity established, adjust the supply fan capacity with its automatic volume control device fully open to provide adequate (but not excessive) static pressure in the primary air duct to the VAV terminal which is most difficult to supply.

.8 Total CFM of the supply air system shall be determined by pitot tube traverse, if practical. If the traverse indicates excessive duct system leakage, this condition shall be recorded in the AABC Report Form.

.9 If not practical to determine the CFM by pitot tube traverse, use the summation of the outlets on each terminal. System must be set in the diversity mode. Do not use the summation of **ALL** the terminals as the supply fan total maximum capacity.

.10 Static pressure shall be measured at the points labeled "P" in Fig. 20.7.

20.13
SECTION IV—SPECIFICATIONS

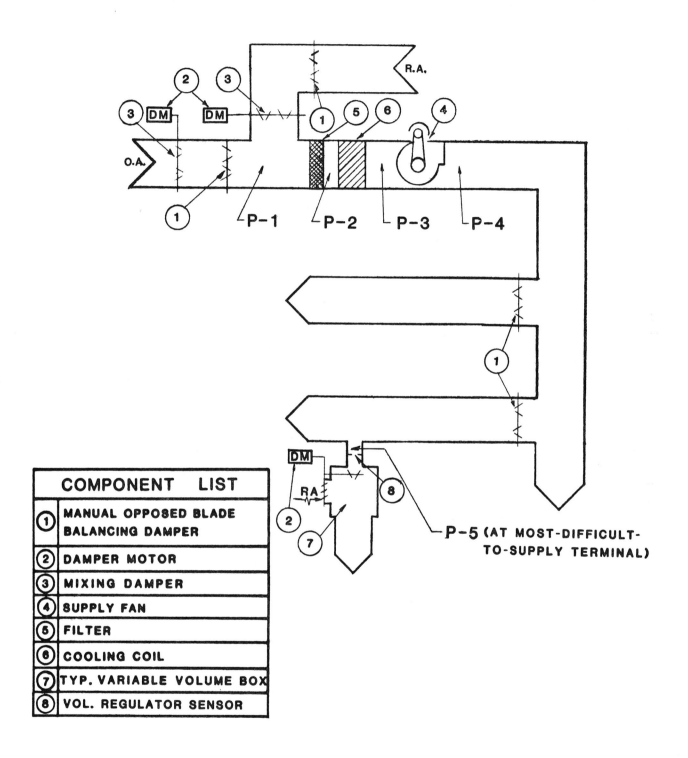

Fig. 20.7: Points at which to measure static pressure in a Variable Primary/Induction Secondary, Pressure Independent VAV System

20.9 VARIABLE PRIMARY/ VARIABLE SECONDARY, SYSTEM POWERED BLADDER TERMINAL SYSTEMS (Fig. 20.8)

.1 The regulator of each VAV terminal shall be set so the indicated value is the same as the design CFM.

.2 Each thermostat shall be cycled from full cooling to full heating to ensure that the VAV terminals (both master and slave units) function as intended.

   A. If the fan is capable of supplying the necessary quantity to meet the total needs of all the terminals, then each terminal must be adjusted in proportion to the available CFM within acceptable tolerances. If the total fan capacity is less than 95%, the individual responsible shall be notified before proceeding with the work.

   B. If the supply fan has an adequate capacity, the constant volume regulator of each terminal shall be adjusted for the required quantity by the AABC Test and Balance Agency.

.3 If boxes do not operate properly and repairs are required, balancing shall be suspended until corrective action is taken.

.4 Set each primary air branch damper so that the downstream static pressure in each branch is sufficient when the system is at maximum flow. The damper in the branch that is most difficult to supply shall be left open.

.5 After all VAV terminals are set and verified for proper functioning, space thermostats shall be set in either a full cooling or heating position as required to satisfy the Design Diversity Factor of the system. Thermostats to be set for heating or cooling shall be selected by the Test and Balance Engineer and shall simulate as nearly as practical the manner in which the system will respond to the cooling load shift of the building.

.6 With proper diversity established, adjust the supply fan capacity with its automatic volume control device fully open to provide the manufacturer's minimum static pressure in the primary air duct to the most remote VAV terminal. Measured static pressure shall be recorded on the AABC Report Form.

.7 Total CFM of the supply air system shall be determined by pitot tube traverse, if practical. If the traverse indicates excessive duct system leakage, this condition shall be recorded in the AABC Report Form.

.8 Static pressure shall be measured at the points labeled "P" in Fig. 20.8.

20.15
SECTION IV—SPECIFICATIONS

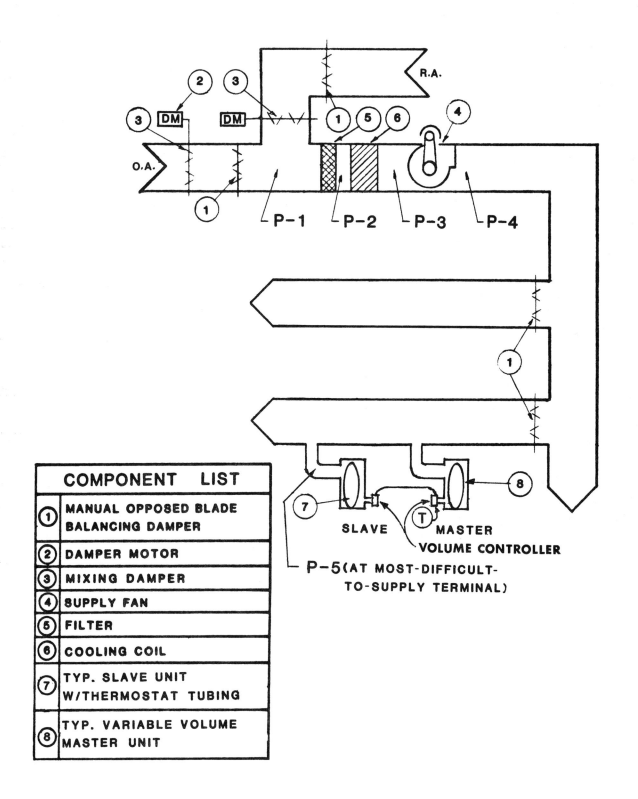

Fig. 20.8: Points at which to measure static pressure in a Variable Primary/Variable Secondary, System Powered Bladder Terminal VAV System

20.10 CONSTANT PRIMARY/ VARIABLE SECONDARY SYSTEMS (Fig. 20.9)

.1 VAV terminals with bypass dampers shall have the manual damper in the bypass positioned so that the bypass quantity is equal to the supply air quantity.

.2 All VAV terminals shall be proportionately balanced to receive the same ratio of primary air. This shall be done by adjusting the manual damper in the inlet duct with all terminals in a full cooling mode (maximum CFM).

.3 At completion of balancing, the inlet manual damper to at least one VAV terminal on each branch duct shall be fully open.

.4 At completion of balancing, at least one damper in each branch duct shall be fully open.

.5 The supply fan shall be set for design air quantity.

.6 The return/exhaust fan capacity shall be set so the building static pressure is slightly positive in a 100% outside air mode. Set the damper in the connection between the return/exhaust fan and the mixed air plenum so the building pressure remains slightly positive in a minimum outside air mode.

.7 Total system CFM shall be determined by pitot tube traverse if practical. If not, the summation of the secondary air outlets shall be used. This information shall be so noted on the AABC Report Forms.

.8 Static pressure shall be measured at the points labeled "P" in Fig. 20.9.

SECTION IV—SPECIFICATIONS

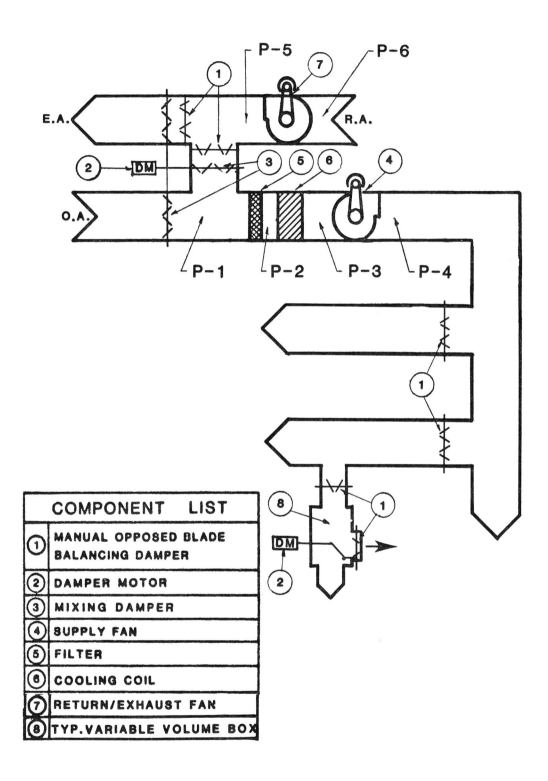

Fig. 20.9: Points at which to measure static pressure in a Constant Primary/Variable Secondary VAV System

20.11 VARIABLE PRIMARY/ VARIABLE SECONDARY, PARALLEL POWERED INDUCTION, PRESSURE INDEPENDENT SYSTEMS (Fig. 20.10)

It is extremely important that the design engineer clearly states the desired sequence of control to be provided for this type system, so that proper verification procedures can be performed by the AABC Test and Balance Agency.

.1 The control system shall be set so that the terminal is operating with no primary air, and with the secondary fan in operation. Adjust the capacity of the secondary fan to the design requirements. Some means must be provided to do this.

.2 The maximum volume regulator of the air valve of each VAV terminal shall be set to supply design CFM with the space thermostat calling for full cooling. There must be adequate static pressure in the primary air duct to overcome the resistance of the secondary ductwork.

.3 After the air valves of all VAV terminals are adjusted, space thermostats shall be set in either a full heating or cooling position as required to satisfy the Design Diversity Factor of the system. Thermostats to be set for heating or cooling shall be selected by the Test and Balance Engineer and shall simulate as nearly as possible the manner in which the system will respond to the cooling load shift of the building.

.4 With proper diversity established, adjust the primary supply fan capacity with its automatic volume control device fully open to provide adequate (but not excessive) static pressure in the branch duct to the VAV terminal air valve which is the most difficult to supply. The measured static pressure shall be recorded.

.5 Set each primary air branch damper so that the downstream static pressure in each branch is sufficient when the system is at maximum flow. The damper in the branch that is most difficult to supply shall be left open.

.6 Total CFM of the primary supply air system shall be determined by pitot tube traverse, if practical. If not practical to determine the CFM by pitot tube traverse, use the summation of the outlets on each terminal. System must be set in the diversity mode. Do not use the summation of **ALL** the terminals as the supply fan total maximum capacity.

.7 Discharge air temperature and CFM shall be measured on each side of all secondary ductwork where tee fittings are installed to determine any stratification. Measurements shall be taken with the terminal in a mode where the cold primary air and warm secondary air streams should mix. Any condition of adverse stratification shall be reported immediately and balancing shall be suspended.

.8 If reheat is provided in a system, the Test and Balance Agency shall verify the proper sequence of operations of all automatic control systems applicable to the reheat system. Automatic control adjustments necessary shall be by others.

.9 Static pressure shall be measured at the points labeled "P" in Fig. 20.10.

20.19
SECTION IV—SPECIFICATIONS

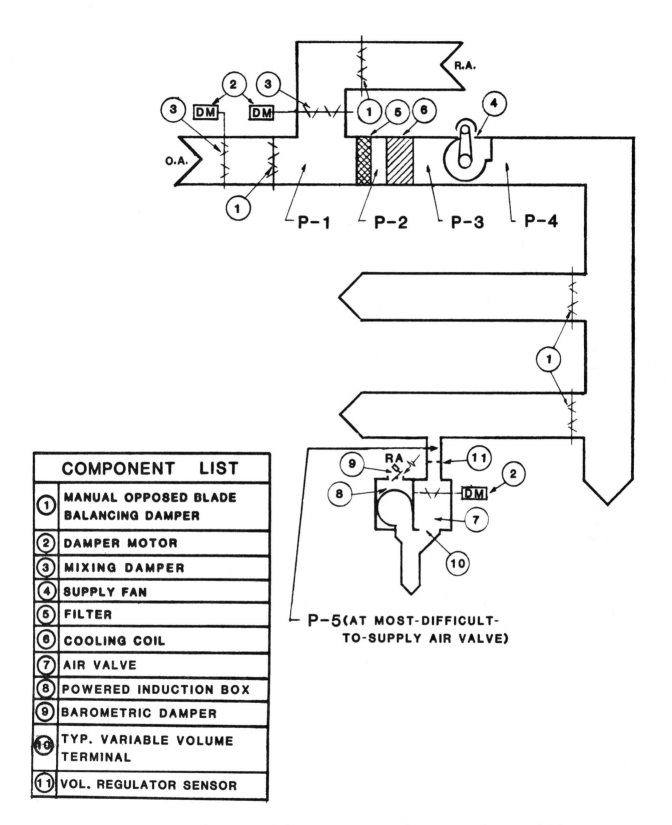

Fig. 20.10: Points at which to measure static pressure in a Variable Primary/Variable Secondary, Parallel Powered Induction, Pressure Independent VAV System

20.12 SUPPLY FAN CONTROLS FOR VARIABLE PRIMARY SYSTEMS

.1 The Test and Balance Agency shall determine the location of the static pressure sensor and the set point of the static pressure regulator which modulates the volume control device on the supply fan. Installation of the sensor shall be by others. Setting of the regulator shall be done by the control contractor in cooperation with the AABC Test and Balance Agency.

.2 The final, established set point of the regulator shall be recorded on the AABC Report Form.

.3 Where an automatic bypass damper is provided for the supply fan, the manual damper in the bypass duct shall be set so the fan delivers essentially the same quantity of air in any mode.

.4 The operation of the variable quantity supply fans shall be observed through a full range of modulation from minimum to maximum CFM to assure that the fan will not be unstable at any point. Space thermostats shall be set to call for heating or cooling as necessary to accomplish this test.

20.13 RETURN/EXHAUST FAN FOR VARIABLE PRIMARY SYSTEMS (Fig. 20.11)

.1 The capacity of the fan shall be set to maintain a slightly positive building static pressure with the following conditions:
   A. Supply air system at maximum quantity.
   B. System at 100% outside air.
   C. Volume control device on the return/exhaust fan fully open.

.2 With the supply air system in a minimum outside air mode and at maximum quantity, the manual damper in the connection between the return fan discharge and mixed air plenum shall be restricted so the building static pressure remains slightly positive. See Fig. 20.11. The volume control device on the return/exhaust fan shall be fully open.

.3 The Test and Balance Agency shall determine that the set point and calibration of the return/exhaust fan variable quantity controls have been properly established.

.4 If a building static pressure controller is used, the Test and Balance Agency shall determine that its location, range and adjustment are correct. The final set point shall be recorded on the AABC Report Form.

.5 If an automatic control scheme is used for tracking the return/exhaust fan with the supply fan, the Test and Balance Agency shall verify the location and adjustment of all devices in the systems.

.6 The operation of the system shall be observed through a full range of modulation from minimum to maximum supply air quantity with the system in both maximum and minimum outside air mode.
   A. It shall be verified that the return/exhaust fan is stable during the full range of modulation.
   B. Mixed air temperatures shall be measured to assure that excessive stratification does not occur which could cause:
      1. Coil freeze-up
      2. Poor coil performance
      3. Non-uniform discharge air temperatures

   In addition, the supply air temperature shall be tested to assure that it will not be below the design value when the system is at minimum outside air mode.
   C. Building static pressure shall be observed during the full range of modulation to verify that it will remain slightly positive at all times.

20.21
SECTION IV—SPECIFICATIONS

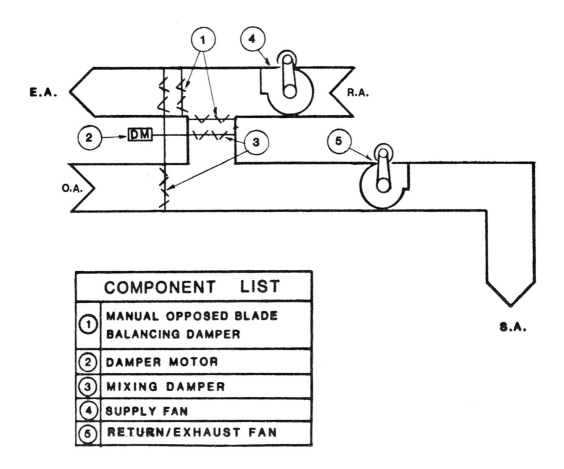

Fig. 20.11: Location of balancing dampers

20.14 RELIEF FAN FOR VARIABLE PRIMARY SYSTEMS

.1 The capacity of the fan shall be set to maintain a slightly positive building static pressure with the following conditions:
   A. Supply air system at maximum quantity.
   B. System at 100% outside air.
   C. The volume control device on the relief air fan fully open.

.2 The Test and Balance Agency shall verify the set point and calibration of controls associated with the relief air fans. These controls shall start and stop the fan and modulate the capacity to maintain a slightly positive building static pressure at all times.

.3 The Test and Balance Agency shall measure the building static pressure in several modes, i.e., 100% outside air, minimum outside air, maximum supply air quantity and minimum supply air quantity. All the measured values shall be recorded on the AABC Test and Balance Report.

Single sheets of specifications suitable for photocopying are available from AABC National Headquarters, upon request.

CHAPTER 21

RETURN AND EXHAUST AIR SYSTEMS

21.1 OVERVIEW

This chapter contains specifications for **Total System Balance** for the procedures common to all return and exhaust air systems.

These specifications include procedures for **Total System Balance** for return and exhaust air inlets and fans.

21.2 PREPARATION FOR TOTAL SYSTEM BALANCE

.1 **Total System Balance** shall not begin until the Test and Balance Agency has verified that the Contractor has completed start-up procedures as listed in these specifications.

.2 The Test and Balance Agency shall measure the motor amperes of all fan motors before **Total System Balance** is begun, and take proper steps to correct and report any overloads before proceeding with the work.

.3 The Test and Balance Agency shall verify all inlets for size, type, and general condition and shall report any variations before starting **Total System Balance**.

21.3 AIR INLETS

.1 All quantities shall be measured according to the AABC National Standards, 1982.

.2 Inlets on systems shall be adjusted to the required quantities with a tolerance of ±10%.

.3 At completion of **Total System Balance**, at least one inlet of every branch shall be fully open and at least one branch balancing damper in the system shall be fully open.

.4 If, during **Total System Balance**, the Test and Balance Agency encounters any conditions that will not allow proper balancing to be performed, the fact shall be reported immediately.

.5 Return air inlets installed in ceilings where the space above the ceiling is used as a return air plenum are not to be measured or adjusted.

21.4 FANS

.1 The Test and Balance Agency shall set the fan RPM to provide design total CFM within acceptable tolerances.

.2 Fan speed shall not exceed the maximum allowable RPM as established by the manufacturer.

.3 The final setting of fan RPM shall not result in overloading the fan motor in any mode of operation.

.4 After **Total System Balance**, the following values shall be measured and recorded:
  A. Fan RPM
  B. Motor voltage and amperes
  C. Static pressure entering the fan (power roof ventilators need not be measured)
  D. Static pressure leaving the fan
  E. Building static pressure with all doors and windows closed.

.5 Static pressure entering and leaving the fan shall be measured as follows:
  A. Static pressure readings leaving the fan shall be taken as far from the fan as is practical, but shall be before any restrictions in the duct (such as duct turns).
  B. No readings shall be taken directly at the fan outlet or through the flexible connection.
  C. Static pressure entering the fan shall be measured in the inlet duct upstream of any flexible connection and downstream of any duct restriction.

D. In all cases, the readings shall be taken to represent as true a value as possible. True value is actual measured static pressure.

.6 Under final balance conditions, the Test and Balance Agency shall measure and record static pressure entering and leaving any heat recovery equipment in the system.

Single sheets of specifications suitable for photocopying are available from AABC National Headquarters, upon request.

CHAPTER 22

HYDRONIC SYSTEMS

22.1 OVERVIEW

Hydronic balancing procedure consists of proportioning fluid flow quantities throughout the system in accordance with design requirements. The Engineer must decide during the project design stage which balancing technique will be used. The techniques include direct reading of meters, indirectly calculating flow rates from pressure differentials across system components, or balancing by a thermal heat balance procedure.

Hydronic system specifications are in a separate chapter from air systems because system design and components are different. However, each AABC Test and Balance Agency accepts the philosophy of **Total System Balance**. This is the belief that one AABC Test and Balance Agency and one Test and Balance Engineer must be responsible for a project if **Total System Balance** is to be achieved. AABC believes that air and hydronic systems are so interdependent that they must be balanced as one integrated system.

22.2 PREPARATION FOR HYDRONIC SYSTEM BALANCING

.1 Hydronic System Balance shall not begin until the AABC Test and Balance Agency has verified the following:
A. System is completely filled.
B. System is clean.
C. System is free of air.
D. All service valves are open.
E. All strainers are provided with clean sleeves having proper perforations.
F. Three-way valves are properly piped.
G. All coils are correctly piped.
H. Coil fins are straight and clean.
I. Proper balancing devices are in place and correctly located:
  1. Meters
  2. Pressure taps
  3. Thermometer wells
  4. Balancing valves
J. Automatic temperature control system is in operation.
K. There is no entrained air in the suction piping to pumps in an open system which can have a negative effect on the pump performance.
L. The pressure is adequate to completely fill the system.

.2 The AABC Test and Balance Agency shall measure the amperes of all pump motors before hydronic balancing is started and shall take proper steps to correct and report any overloads.

.3 The AABC Test and Balance Agency shall not continue the hydronic balancing if at any time hazardous conditions are observed. These conditions shall be reported before proceeding further.

22.3 GENERAL PROCEDURES

.1 All flow quantities, temperatures and pressures shall be measured according to the AABC National Standards, 1982.

.2 If, during the hydronic balancing, the AABC Test and Balance Agency determines any conditions that will not permit proper balancing, the fact shall be reported immediately.

.3 At completion of balancing, at least one terminal unit balancing valve in each piping branch shall be fully open.

Chapter 22—Hydronic Systems

.4 At completion of balancing, at least one branch pipe balancing valve shall be fully open.

.5 The final position of each balancing valve shall be clearly marked. Any memory devices shall be set to permit closing and reopening the valve to its balanced setting.

.6 The systems shall be balanced so the flow tolerance is in accordance with Fig. 22.1. To use this nomograph, draw a horizontal line through the supply water temperature on the left to the appropriate diagonal Delta T line. Project vertically, upward or downward, to determine the allowable tolerance.

.7 The AABC Test and Balance Agency shall verify that all automatic controllers operate the correct control valves. The valve position shall be as indicated by the controller.

22.4 FLOW METER BALANCE PROCEDURE

.1 Fluid flow quantities shall be measured using the installed meters provided by others.

.2 The AABC Test and Balance Agency shall apply any necessary correction factor to the indicated value to account for the density of the fluid flowing in the system.

.3 The initial and final readings of all meters shall be included on the AABC Report Form. All pertinent information regarding each meter shall be listed, such as:
  A. Designation of terminal
  B. Manufacturer
  C. Type
  D. Size
  E. Rating
    1. GPM
    2. Pressure differential

.4 If specified, pitot tube traverses shall be taken where required on the drawings, provided valved openings are properly installed.

22.5 SYSTEM COMPONENT BALANCE PROCEDURE

.1 General Discussion

Balancing by flow meters will be much more accurate than by using pressure differential across system components. However, if the only available measuring points are pressure differential across system components, such as:
  A. Coils
  B. Temperature control valves
  C. Heat exchangers
  D. Chillers
  E. Other terminals

Then such components may only be used to approximate the flow.

The validity of any calculated flow rates from field measured pressure differential will depend upon the reliability of the manufacturer's data. The accuracy of the flow rates determined by this procedure may be beyond the allowable tolerances indicated by Fig. 22.1. It is imperative that the pressure taps used for these field measurements be located as close to the components as possible. They should not be located where entering or leaving conditions would not give true static pressure indications.

.2 Specification

A. Fluid flow quantities shall be calculated by using the measured differential pressure across the system components and comparing it with the manufacturer's flow vs. pressure differential rating. The following equation shall be used:

$$\frac{GPM_M}{GPM_R} = \sqrt{\frac{\Delta P_M}{\Delta P_R}}$$

GPM_M = Measured GPM
GPM_R = Rated GPM
ΔP_M = Measured Pressure Differential
ΔP_R = Rated Pressure Differential

B. The AABC Test and Balance Agency shall apply any necessary correction factor to the indicated value to account for density of the fluid flowing in the system.

C. The initial and final values of all stations (components used as flow meters) shall be included on the AABC Report Form. All other pertinent information shall be listed, such as:
   1. Designation of station
   2. Rated GPM
   3. Rated pressure differential

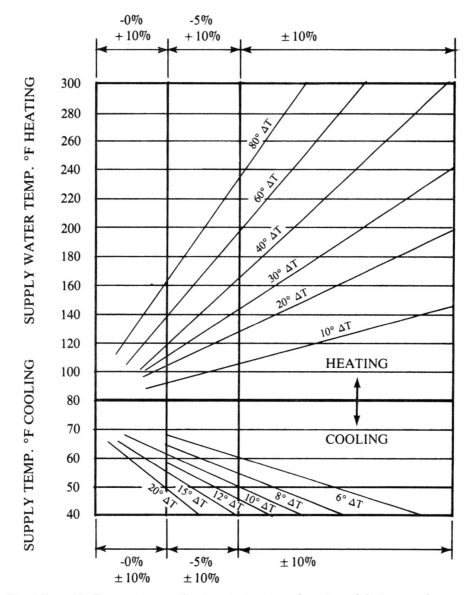

The Allowable flow tolerance (final to design) as a function of design supply water temperature and design water temperature differential.

Note: Ability to obtain these tolerances depends on the accuracy of the meter installation.

(Note: Modification of ASHRAE, 1980 Systems Handbook, page 40.8, Fig. 4)

Fig 22.1: Flow tolerance plot

22.6 TEMPERATURE DIFFERENCE (THERMAL) BALANCE PROCEDURE

.1 General Discussion

A. Thermal balancing is not a recommended procedure where a high degree of accuracy is required. It is not recommended in the following systems:
   1. Low temperature hot water heating system.
   2. Hot water heating system with a high design differential water temperature.
   3. Hot water heating system containing different types of components.
   4. Hot water heating system where components have varying required temperature differentials.
   5. Any cooling system.

B. Where thermal balance procedure is required, the system can be balanced in the manner described in this section.

C. Fundamentally, thermal balancing consists of calculating fluid flow quantities by heat balance.

D. Hot water heating systems using supply water temperatures in the range of 200°F with not over 40°F rated temperature differential are possible candidates for this procedure where radiation or unit heaters are installed. To be performed with expediency, the rated water side temperature differential for all terminals should be the same. This allows the Technician to balance the system by causing the return water temperature of each terminal to be identical provided the supply water temperature is constant.

E. If terminals with different rated temperature differential are provided, the task is more difficult. An approximation of the ratio of measured to rated flow can be determined by using the equation:

$$\frac{GPM_C}{GPM_R} = \frac{\Delta T_R}{\Delta T_M}$$

$\frac{GPM_C}{GPM_R}$ = Ratio of Calculated GPM to Rated GPM

ΔT_R = Rated Water Temperature Differential

ΔT_M = Measured Water Temperature Differential

This assumes the supply water temperature is as rated and that the terminals are radiation, unit heaters or other heating equipment not associated with the supply air system. However, for all practical purposes, proportionate thermal balancing can be performed at other than rated water temperatures using the above equation provided the supply water temperature is constant during the balancing procedure.

F. Where hot water heating coils, or chilled water cooling coils with no latent load, are provided, the equation in Section 22.6.1E can be used for proportionate balancing of the hydronic system provided the airflow quantities across the coils have been balanced, and provided also, that the entering air temperatures are the same for each coil.

G. Obviously, it becomes much more difficult, if not in fact impossible, to thermally balance a mixture of heating coils, unit heaters and/or radiation. Where such conditions occur and the Engineer has determined that the balancing must be done thermally, it will be necessary to calculate GPM flow rate by use of the following equation:

$$GPM = \frac{TH}{\Delta T \times 500}$$

GPM = Gallons Per Minute
TH = Total Heat Transferred (BTUH)
ΔT = Measured Water Temperature Differential

If the CFM of a coil is known, the air temperature differential across the coil can be measured at one point entering and one point leaving the coil. This will approximate Total Heat Transferred so the GPM can be estimated. Calculating the GPM flow rate through radiation or unit heaters is essentially impractical in the field unless flow meters are provided.

H. Where latent heat removal is involved, thermal balancing should not be attempted. The need for accurate results, plus the additional time to measure both dry bulb and wet bulb temperatures, and to calculate total heat to estimate GPM flow rate makes the use of flow meters imperative.

I. The following specification section reflects the AABC philosophy of thermal balance procedure.

.2 Specification

Hot Water Heating System with All Terminals of Same Design Entering and Leaving Water Temperatures and No Air Balance Required.

A. Balancing valves shall be set so that the measured return water temperature at each terminal is identical.

B. The supply water temperature during the balancing procedure must be constant (not necessary to be design temperature).

C. A proportional balance will be established when the temperature tolerance is not greater than ±10% as established by the following equation.

$$\% \text{ Balance} = \frac{\Delta T(Av)}{\Delta T(Final)}$$

$\Delta T(Av)$ = Average of (EWT − LWT) of all Terminals
$\Delta T(Final)$ = Final Measured (EWT − LWT) of all Terminals

C. The initial and final water temperature readings of all terminals shall be included in the report. Include designation of each terminal.

Hot Water Heating System with All Terminals of Same Design Entering and Leaving Water Temperatures—Air Balance Required.

A. Balancing valves shall be set so the measured return water temperature at each coil is identical.

B. The supply air system must have been balanced prior to the hydronic balancing.

C. The supply air temperature and entering water temperature shall be maintained constant during the balancing procedure.

D. A proportional balance will be achieved when the temperature tolerance is not greater than ±10% as established by the following equation:

$$\text{Ratio} = \frac{\Delta T(Av)}{\Delta T(Final)}$$

$\Delta T(Av)$ = Average (EWT - LWT) of all Coil
$\Delta T(Final)$ = Final Measured (EWT - LWT) of the Coil

Hot Water Heating System, with All Terminals of Same Design Entering Water Temperature and Different Leaving Water Temperature—No Air Balance Required.

A. Balancing valves shall be set so the ratio of measured ΔT to rated ΔT of all coils is identical.
B. The supply water during the balancing procedure must be maintained at a constant temperature.
C. A proportionate balance will be established when the ratio tolerance is not greater than ± 10%.

$$\text{\% Balance} = \frac{\text{Ratio (Av)}}{\text{Ratio (Final)}}$$

$$\text{Ratio (Av)} = \text{Average Ratio of } \frac{\text{Rated } \Delta T}{\text{Measured } \Delta T} \text{ of all Terminals}$$

$$\text{Ratio (Final)} = \text{Final Measured Ratio of } \frac{\text{Rated } \Delta T}{\text{Measured } \Delta T} \text{ of the Terminal}$$

Hot Water Heating System with All Terminals of Same Design Entering Water Temperature and Different Leaving Water Temperature—Air Balancing Required.

A. Balancing valves shall be set so the ratio of measured ΔT to rated ΔT of all coils is identical.
B. The supply air system must have been balanced prior to the Hydronic Balancing.
C. The supply air temperature and entering water temperature shall be maintained constant during the balancing procedure.
D. A proportionate balance will be established when the ratio tolerance is not greater than ± 10%.

$$\text{\% Balance} = \frac{\text{Ratio (Av)}}{\text{Ratio (Final)}}$$

$$\text{Ratio (Av)} = \text{Average Ratio of } \frac{\text{Rated } \Delta T}{\text{Measured } \Delta T} \text{ of all Coils}$$

$$\text{Ratio (Final)} = \text{Final Measured Ratio of } \frac{\text{Rated } \Delta T}{\text{Measured } \Delta T} \text{ of the Coil}$$

22.7 SELF-CONTAINED, AUTOMATIC FLOW LIMITING DEVICE— BALANCING PROCEDURES

.1 General Discussion

The installation of self-contained, automatic flow limiting devices in a hydronic system does not make it self-balancing. In order to control at rated flow, the differential pressure across the device must be within the rated pressure differential limits at all times. They do not proportion the system. They merely control the flow to a preset maximum value provided there is sufficient pressure differential to allow them to control.

If the pressure drop in any branch is in excess of the upper limit of the control range of the controlling device, the flow limiting device will be out of control and a manual balancing valve must be used.

If the pressure drop is less than the lower limits of the controlling devices in the branch, the flow limiting device will be out of control and one valve in other branches must be restricted in order to proportion the flow through the system.

In any event, provisions should be made for ease in removing the flow limiting device for cleaning. Pressure taps should be provided to allow the AABC Test and Balance Agency to verify that there is enough pressure differential to insure that the regulator is in control.

.2 Specification

The AABC Test and Balance Agency shall verify that all flow limiting devices have a rated capacity to meet the design intent. It shall be verified that the device is installed properly.

The pressure differential shall be measured across each device. If the pressure differential is found to be inadequate or excessive, the balancing procedure shall be stopped until corrective action is taken. The AABC Test and Balance Agency shall verify:
A. That temperature control valves are operated by the intended controller.
B. The integrity of temperature control valves.
C. That valves are correctly installed.
D. That elements are correctly piped.
E. That all service valves are open.

The following information shall be included in the AABC Report Form for each terminal with a flow limiting device:
A. Terminal designation
B. Design GPM
C. Self-contained, automatic flow limiting Device Data
  1. Manufacturer
  2. Catalog number
  3. Rated GPM
  4. Rated differential pressure
  5. Measured differential pressure
  6. Size

22.8 CONSTANT FLOW SYSTEMS

.1 Where systems are provided with three-way temperature control valves, the balancing valve in the bypass connection shall be restricted so that the flow rate will be not more than 60% of design flow rate when in a full bypass mode.*

.2 After balancing is completed, the pump discharge throttling valve shall be set so each terminal receives the rated flow quantity.

Alternate Procedure (To reduce energy consumption)

The AABC Test and Balance Agency shall calculate the new pump impeller diameter that would be required to provide the design GPM with the discharge throttling valve fully open. The proper size impeller will be furnished and installed by others. After the modified impeller is installed in the pump the AABC Test

---

*Note to Engineer: See Chapter 14 of these AABC National Standards, 1982 for piping arrangements, location of measuring devices, and pressure taps.

and Balance Agency shall remeasure the following:
- Total GPM
- Discharge static pressure
- Suction static pressure
- Motor amperes

.3 After **Total System Balance,** the following values shall be recorded:
- Motor voltage and amperes
- Discharge static pressure
- Suction static pressure
- Block tight head

22.9 VARIABLE FLOW SYSTEM

.1 Sufficient valves shall be opened or closed to simulate design diversity if applicable.

.2 All bypass valves shall be set.

.3 The pump discharge throttling valve shall be set so each terminal receives rated flow quantity.

Alternate Procedure (To reduce energy consumption)

The AABC Test and Balance Agency shall calculate the new pump impeller diameter that would be required to provide the design GPM with the discharge throttling valve fully open. The proper size impeller will be furnished and installed by others. After the modified impeller is installed in the pump, the AABC Test and Balance Agency shall remeasure the following:
- Total GPM
- Discharge static pressure
- Suction static pressure
- Motor amperes

.4 After **Total System Balance,** the following values shall be recorded:
- Motor voltage and amperes
- Discharge static pressure
- Suction static pressure
- Block tighy head

22.10 PUMPS

.1 Where there are no meters in the system, the pump shall be used for estimating the total system flow rate. See Chapter 9 of the AABC National Standards, 1982.

.2 Where parallel pump operation is provided, the motor amperes shall be measured with one pump operating to insure there is no overload.

22.11 COIL BANKS

.1 To compensate for any stratification of air temperature or uneven air velocity across the coil bank, the water flow through banks of multiple coil sections shall be balanced thermally so that the return water temperature of each coil is the same. The balancing valve of at least one coil shall be fully open.

22.12 COOLING TOWER

.1 This specification is not for a Cooling Tower Performance Test. See Chapter 23 of the AABC National Standards, 1982.

.2 Where multiple distribution pans are provided, the water in all the pans of the upper basin shall be set to the same level.

.3 Where a multi-cell tower is provided with separated lower basins, the water in the lower basins shall be set to the same level.

.4 Where a three-way automatic control diverting valve is provided, the balancing valve in the bypass shall be set to provide constant water flow rate through the condenser.

Single sheets of specifications suitable for photocopying are available from AABC National Headquarters, upon request.

CHAPTER 23

SPECIAL SYSTEMS

23.1 OVERVIEW

This chapter contains testing specifications for those special systems that are not usually contained in specifications, but which are accepted as the responsibility of the Associated Air Balance Council. Since these are not normally included in the specifications for environmental systems, they must be specifically requested.

The AABC Test and Balance Engineer is qualified in all areas of **Total System Balance**, including the work described in this chapter; as proven by meeting the AABC requirements for Certification. The work described requires special skills, knowledge, and instrumentation. All AABC Agencies are certified for this work and possess the needed instrumentation.

This chapter contains testing specifications for the following:
23.2 Ceiling Plenum Systems
23.3 Sound
23.4 Vibration
23.5 Combustion
23.6 Cooling Tower
23.7 Duct Leakage
23.8 Fume Hoods

23.2 CEILING PLENUM SYSTEMS

Before a ceiling plenum system can be balanced, it is necessary to have the correct CFM supplied to the plenum. If design CFM is not available, proper room penetration may not be achieved.

The AABC Test and Balance Agency shall:
.1 Before the ceiling is installed, make a visual inspection, using a high intensity light source, for potential leakage at barriers, seals, and at all other possible locations.
.2 Test and report any air leakage through ceiling tile joints and around the ceiling perimeter.
.3 Test and record static pressure in the ceiling plenum.
.4 Test and record air velocity on a grid of ten foot centers, at six feet above the floor.
.5 Adjust ceiling tile perforations, if provided, to achieve a uniform air terminal velocity in the room at between 25 to 35 FPM, at six feet above the floor and measured on a grid of ten foot centers. If lay-in tiles are provided, rearrange as necessary to achieve the same results.

23.3 SOUND TESTING

.1 Overview

This Section contains the specifications for sound pressure level testing. It includes the responsibilities of the Testing and Balancing Agency, the Engineer, and others.

A. Responsibilities of the Testing and Balancing Agency are:
- Measure and record sound pressure readings as specified.
- Plot NC curves, if specified.
- Report any unusual conditions at time of test.
- Identify noise source in any non-complying situation.

B. Responsibilities of the Engineer
The responsibilities of the Engineer are:
- Specify the allowable sound pressure levels.
- Clearly specify the areas to be tested.
- Specify the typical spaces within the area to be tested.

C. Responsibilities of others

The responsibilities of others are:
- Analyze the readings taken by the Testing and Balancing Agency.
- Take any needed corrective action.
- Payment for any additional testing required by failure of initial test.

D. There are two acceptable approaches to sound pressure test specifications according to the type of building and type of occupancy:
1. Only dbA Levels specified.
2. Octave Band Analysis specified.

.2 General

A. Sound testing shall be performed by a Test and Balance Agency certified by the Associated Air Balance Council (AABC). All work done by this Test and Balance Agency shall be by qualified Technicians under the direct supervision of an AABC Certified Test and Balance Engineer.

B. Sound testing shall be performed in complete accordance with the AABC National Standards, 1982.

C. No interior sound measurements shall be taken before a building is completely enclosed to eliminate outside sound. All testing and balancing shall be completed.

.3 Locations

The Test and Balance Agency shall take sound measurements in the following specific locations:

A. _____
B. _____
C. _____
D. _____
E. _____

.4 Submittals

A. The Test and Balance Agency shall submit all recorded sound measurements and data on approved AABC Report Forms. These reports shall include the following for each tested location:
1. Location of test.
2. Alternate Specification: dbA Levels. dbA scale readings:
   a. With equipment being tested turned off.
   b. With equipment being tested turned on.
3. Alternate Specification: Octave Band Analysis.
   a. Take readings at each of the following 8 octave band center frequencies—63, 125, 250, 500, 1000, 2000, 4000, 8,000 and plot an NC curve to determine the NC level:
      1. With equipment being tested turned off
      2. With equipment being tested turned on
   b. The NC level at each tested location determined by plotting.
      1. With equipment on
      2. With equipment off
4. The model and type of sound level meter, octave band analyzer, microphone, and calibrator used.

.5 Measurements

A. The AABC Test and Balance Agency shall make all preparations for measurement, the measurements themselves, and the record of the measurements according to the procedures specified in Chapter 12 of the AABC National Standards, 1982.

.6 Interior Sound

A. Alternate Specification: dbA Levels
1. The AABC Test and Balance Agency shall for each location specified take a series of readings (with equipment on and equipment off) in accordance to the methods specified in the

AABC National Standards, 1982, Chapter 12.
2. If the dbA readings exceed the value specified, new readings shall be taken on the B scale and the C scale (with equipment on and off). These new readings are to be recorded on the Report Form.

B. Alternate Specification: Octave Band Analysis
1. Take readings at each of the following 8 octave band center frequencies—63, 125, 250, 500, 1000, 2000, 4000, 8000, and plot an NC curve to determine the NC level.
   a. With equipment being tested turned off
   b. With equipment being tested turned on
2. If the NC level exceeds the specified NC curve, turn off the equipment, one piece at a time to identify the source of the disturbing sound and plot an NC curve

.7 Exterior Sound

In measuring exterior sound, the AABC Test and Balance Agency shall:
A. Measure all exterior sound pressure levels in dbA at the specified location.
   1. With equipment on
   2. With equipment off
B. Testing shall be done at a time of minimum background sound.
C. Exterior sound measurements shall not be taken if more than one inch of snow is present.

23.4 VIBRATION TESTING
.1 Overview

This Section contains the specifications for vibration level testing. It includes the responsibilities of the Testing and Balancing Agency, the Engineer, and others.

A. Responsibilities of the Testing and Balancing Agency
The Responsibilities of the Testing and Balancing Agency are:
- Measure and record vibration level readings as specified.
- Report any unusual conditions at time of test.
- Identify vibration source in any non-complying situation.

B. Responsibilities of the Engineer
The responsibilities of the Engineer are:
- Specify the allowable vibration levels.
- Clearly specify the equipment to be tested.

C. Responsibilities of others
The responsibilities of others are:
- Analyze the readings taken by the Testing and Balancing Agency.
- Take any needed corrective action.
- Payment for any additional testing required by failure of initial test.

D. The purpose of vibration testing is to determine whether equipment in environmental systems is subject to sufficient vibration to result in damage or shortened equipment life.

E. Vibration testing should only be specified on occasions when there is reason to suspect excessive vibration in environmental system equipment.

F. When vibration testing is requested, the specific items of equipment to be tested should be carefully specified.

.2 General Requirements
A. Prior to vibration testing, the AABC Test and Balance Agency shall inspect and determine if the vibration isolators on equipment meet specifications by type and load ratings and are installed correctly.

.3 Scope of Work
A. Vibration testing shall be performed by an AABC Test and Balance Agency cer-

tified by the Associated Air Balance Council (AABC). All work done by this AABC Test and Balance Agency shall be by qualified Technicians under the direct supervision of an AABC Certified Test and Balance Engineer.
B. Vibration testing shall be performed in complete accordance with the AABC National Standards, 1982, Chapter 13.
C. No vibration measurements shall be taken before all systems are tested and balanced.
D. The AABC Test and Balance Agency shall measure the vibration in each specified piece of equipment for velocity and displacement. This data shall be recorded on the AABC Report Form. All measurements shall be taken under actual operating conditions.
E. For the equipment specified, readings shall be taken on each bearing cap in vertical, horizontal and axial directions. If bearings are inaccessible, the same readings shall be taken at the equipment mounting feet.
F. The Test and Balance Agency shall make one test only on each piece of specified equipment.

.4 Submittals
A. The AABC Test and Balance Agency shall submit all recorded vibration measurements and data on approved AABC Report Forms. These AABC Reports shall include the following for each testing location:
 1. Equipment Designation
 2. Motor HP
 3. Location of pickup
   a. Vertical
   b. Horizontal
   c. Axial
 4. Velocity and Deflection
 5. Equipment RPM

23.5 COMBUSTION TESTING
.1 Overview
A. The primary purpose of combustion testing is to determine flue gas loss of fossil-fueled equipment, such as boilers and furnaces. The secondary purpose is to determine gross MBH (Thousand BTU per Hour) input and gross MBH output.
B. The intent of combustion testing as performed by an approved AABC Test and Balance Agency is to **test only.** The intent **is not** to make any adjustments in air-fuel ratios. This is the responsibility of combustion specialists.
C. Combustion Testing is not normally included in Environmental System specifications and must be specifically requested.
D. The Engineer must specify any options desired under 23.5.2E—Optional Testing. Also, the proper fuel specifications must be selected.
E. Payment for additional tests required shall be by others.

.2 Specifications
A. All measured and collected data shall be recorded on the approved AABC Report Form.
B. The following non-test data shall be recorded:
 1. Equipment designation number
 2. Equipment manufacturer
 3. Model number
 4. Serial number
 5. Rated input
 6. Rated output
 7. Type of fuel, and heat value (calorific value)
 8. Other pertinent data.
C. The following measurements shall be taken and recorded to meet minimum requirements:
 1. Percent CO_2

2. Stack temperature (flue gas temperature at equipment outlet)
3. Ambient temperature
D. The Testing and Balancing Agency shall make one test only on each piece of specified equipment.
E. Optional Testing
The following additional measurements shall be taken and recorded:
1. Percent CO
2. Percent O_2
3. Over-fire draft (inches WG)
4. Gas burner manifold pressure
5. Fuel meter reading at burner, (if meter is provided)
6. Boiler jacket temperature

.3 Instrumentation
A. Any portable combustion analyzer is acceptable for the measurements specified. Orsat apparatus is not required.

.4 Measurements (Fig. 23.5.1)
A. If burners have other than one firing rate, combustion tests shall be made in the following manner:
1. If step-firing (High-Low), readings shall be taken at each step.
2. If modulating, reading shall be taken at high, at low, and at 50% firing rate.
B. Fuel Measurements
1. Natural gas
   a. The objective of this test is to determine gross MBH input.
   b. Time the gas meter under each firing rate to determine the CFH (Cubic Feet per Hour) delivered. CFH shall be calculated by the equation in Fig. 23.5.1.
   C. Measure or determine the gas pressure at the gas meter and apply the proper correction factor to the delivered CFH that was measured to determine the equivalent CFH at the rated gas pressure. This shall be determined by the following equations:

$$F = \frac{P_A + P_G}{P_A + P_R}$$

F = Correction Factor
P_A = Atmospheric Pressure (psi)
P_G = Gas Pressure at Meter (psi)
P_R = Rated Gas Pressure (psi)

Rated CFH = Measured CFH × Correction Factor

2. Oil Fired
   a. The oil pressure at the pump shall be measured and recorded. Oil flow rate shall be determined if manufacturer's data is available.
3. LP Gas Fired
   a. The burner manifold pressure shall be measured. Gas flow rate shall be determined if manufacturer's data is available.

.5 Calculations
The following calculations shall be made from the measured data and recorded on the Report.
A. Gross Input (MBH)
B. Net Stack Temperature
C. Percent Excess Air
D. Percent Flue Gas Loss
E. Combustion Efficiency
F. Gross Output (MBH) = Gross Input − Flue Gas Loss

Flue gas loss shall be calculated from measured data using industry accepted methods.

$$\text{CFH} = \frac{\text{Value of timing dial for one rotation} \times 3600}{\text{Time in seconds for one rotation of timing dial}}$$

Fig. 23.5.1: Equation for Cubic Feet per Hour

23.6 COOLING TOWER CAPACITY PERFORMANCE TEST

.1 Overview

This section covers Cooling Tower Capacity Performance **Test only**.

The cooling tower is one of the most important parts of the air conditioning system. Therefore, as part of **Total System Balance**, the cooling tower should be tested to determine its operating capacities, as installed. This installed operating capacity must then be compared to the rated performance of the tower. The resulting Report quickly tells the Engineer that the installed cooling tower will satisfy the heat rejection requirement from the refrigeration unit at design conditions. It is the responsibility of the AABC Test and Balance Agency to take the specified tests and to record the results on the Report. The Test and Balance Agency shall make one test only on each piece of specified equipment. It is the responsibility of others to designate the time of tests.

Measurements for cooling tower capacity performance must be taken in a manner to assure an unusually high degree of accuracy. Otherwise, the resulting Report will have little meaning. To achieve the required accuracy, **it is essential** that the AABC Test and Balance Agency designate the locations of test points in the system, and the type of meters to be used. These test points and meters are to be installed by others at the time the system is installed.

Calculations for this Report are to be done in accordance with AABC Technical Report, "Cooling Tower Field Testing and Conversion Data." See the Appendix of these AABC National Standards, 1982.

.2 Specification

A. Cooling Tower Capacity Performance Tests shall be performed by an Agency certified by the Associated Air Balance Council (AABC).

B. Tests and calculations shall be performed in accordance with AABC Technical Report, "Cooling Tower Field Testing and Conversion Data." See the Appendix of these AABC National Standards, 1982.

C. Measurements shall be taken as specified in this Section, and also as specified in the AABC National Standards, 1982, Section II.

.3 Work of Other Trades

A. The water circulating system that serves the cooling tower shall have been thoroughly cleaned of all foreign matter. Samples of water shall be clear and indicate clear passage of water through pumps, piping and screens.

B. Fans serving the cooling tower shall be rotating correctly. All obstructions shall have been removed from the path of air flow. Permanent obstructions shall be noted by the AABC Test and Balance Agency.

C. Interior filling of cooling tower shall be clean and free of foreign materials such as scale, algae or tars.

D. The water in the upper and lower basins of the tower shall be set at proper working levels.

E. Water circulating pumps shall have been tested for correct operation and placed in normal operating conditions.

.4 General Procedures

A. The AABC Test and Balance Agency shall check the lower basin during full flow to determine that the centrifugal or vortex action of the water is not causing entrainment of the air.

B. All measurements shall be made in as short a time period as possible so that nearly simultaneous readings are observed.

C. The AABC Test and Balance Agency shall use a team of no less than one Test

SECTION IV—SPECIFICATIONS

and Balance Engineer and one Technician to perform this test in order to achieve as nearly simultaneous readings as possible.
D. The test shall be run, recorded, and plotted at least two times. If the plotted results do not agree, the test shall be repeated until two plotted curves overlay within 5% of each other.
E. Inlet and outlet air temperature readings shall be made using the prescribed instruments in accordance with the AABC National Standards, 1982.
F. Conditions at the time of test shall be:
 1. Water flow rate within 15% of design.
 2. Heat load within 30% of design and constant during the test.
 3. Entering wet bulb within 12°F of design wet bulb.
 4. The blowdown (bleed off) is closed.
 5. The make-up water valve is closed.
 6. Throttle valve at the pump discharge is set to provide the full rated flow across the cooling tower. This flow must be maintained constant during the entire test.
G. Water Temperature Readings
 1. All temperature readings are to be taken only by the type of thermometer specified.
 2. Condenser water supply (cool water) temperatures shall be read by placing the thermometer in a receptacle and draining system water from the blowdown valve into the receptacle.
 3. Condenser water return (hot water) temperatures shall be read by placing the thermometer in the upper basin. In a closed cooling tower, an appropriate thermometer well, as hereinafter specified, shall be installed where designated by the AABC Test and Balance Agency.
H. Air wet bulb temperatures shall be taken with a powered psychrometer at the following locations.
 1. Fan discharge.
 2. Intake air.
 3. Ambient, away from the influence of the tower.
I. Intake air wet bulb temperature readings shall be taken as follows:
 1. All readings shall be taken at a maximum of three feet from the tower.
 2. Readings shall be taken on a grid of approximately five foot centers, and the average of these readings calculated.
J. No dry bulb temperatures and no fan CFM readings shall be taken.

.5 Instrumentation

A. Thermometers shall be test-quality, glass-stemmed, mercury filled, 0°F to 120°F, with an etched scale in 0.2°F subdivisions.
B. All thermometers used in the test shall be checked against each other to assure that they have identical readings.
C. Thermometer wells shall be the type that accepts a glass-stemmed test thermometer of the type specified in this section. The well shall be filled with a heat transfer media during the readings.
D. Only the following are acceptable for flow measurements:
 1. Calibrated orifice.
 2. Venturi.
 3. Pitot tube traverse.
E. Only the following gages are acceptable, see Chapter 8 of the AABC National Standards, 1982.
 1. Manometer.
 2. Hydronic Differential Gage.
F. Flow quantities determined by pressure differential across system components, or by pump pressure rise will not be acceptable.

.6 Submittals

A. Data recorded on the Report shall include the following as minimum.
 1. Data specified in the AABC National Standards, 1982, Chapter 26.

23.8 Chapter 23—Special Systems

2. Ambient wet bulb temperature.
3. Barometric pressure.
4. Wind velocity and direction.
5. Remarks of any unusual conditions affecting tower performance, such as obstructions to airflow; air recirculation; or tower condition.

B. A cooling tower performance curve shall be plotted and attached to the Report.
1. The horizontal coordinate shall be heat rejected, in 1000 BTU/minute.
2. The vertical coordinate shall be for Approach in degrees F.
3. The design condition shall also be plotted and identified.
4. The specified approach shall be plotted and identified.
5. The heat rejected at design approach shall be plotted and identified.

C. As part of the Report, the ratio of Test Heat Rejected to Design Heat Rejected shall be calculated and stated as, "Cooling Tower Percent of Design Capacity."

23.7 DUCT LEAKAGE TESTING
.1 Overview

Excessive duct leakage will make **Total System Balance** ineffective. A 10% duct leakage can result in a 30% greater horsepower requirement for the fan.

Since the duct system is tested in sections, the leakage shall not exceed $\frac{1}{2}$ of 1% of the CFM to be handled by that section, and the total leakage of the system shall not exceed 1% of the total system CFM. However, the test pressure should never exceed the pressure limits of the duct construction as defined in SMACNA's "High Pressure Duct Construction Standards."

This Section contains the specifications for duct leakage testing. It includes the responsibilities of the Testing and Balancing Agency, the Engineer, and others.

A. Responsibilities of the Testing and Balancing Agency are:
* Measure and record duct leakage rate as specified.
* Report any unusual conditions at time of test.
* Identify leakage source in any non-complying situation.

B. Responsibilities of the Engineer
The responsibilities of the Engineer are:
* Arrange to witness any desired test.
* Specify the allowable duct leakage rate.
* Clearly specify the systems to be tested.

C. Responsibilities of Others
The responsibilities of others are:
* Analyze the readings taken by the Testing and Balancing Agency.
* Take any needed corrective action.
* Payment for any additional testing required by failure of initial test.

It is not recommended that low pressure duct systems be leak tested. However, low pressure systems should be sealed. Leakage of low pressure systems creates a substantial problem when Testing and Balancing. If these systems are sealed in accordance with SMACNA Low Pressure Duct Construction Standards, the leakage rate should not be over 5% of total system CFM.

.2 Specifications

A. All openings in the duct system shall be carefully sealed shut by others. Sealing shall conform to the recommendations of the AABC Test and Balance Agency.
B. Duct systems shall be separated and sealed in sections according to the recommendations of the AABC Test and Balance Agency.

SECTION IV—SPECIFICATIONS

.3 Testing Procedure
(Fig. 23.7.1; 23.7.2)

A. Testing shall be by orifice tube as shown in Figs. 23.7.1 and 23.7.2. The test kit shall consist of:
1. Test blower.
2. Calibrated orifice tube.
3. Two manometers.
4. Flexible tubing for connecting to the duct system.

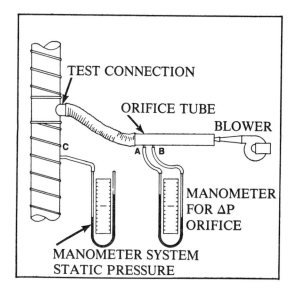

Fig. 23.7.2: Leakage testing by orifice tube*

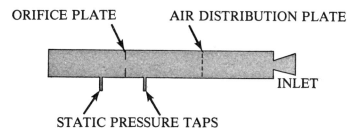

Fig. 23.7.1: Orifice tube*

B. The test apparatus shall be connected to the test connection in the system (Fig. 23.7.2). The system static pressure tap (C) shall be in the system duct and shall be at least 12″ from the test connection. Static pressure taps in the orifice tube **shall not** be used to read system static pressure.

C. The inlet opening of the test blower shall be blocked off before the test blower is started. The inlet opening shall then be opened slowly to prevent overpressurizing the system.

D. Flow across the orifice shall be determined by the following equation (for standard air).

$$Q = 4005 \, CA \sqrt{\Delta P}$$

Q = CFM
C = Orifice coefficient
A = Area of orifice in square feet
ΔP = Pressure difference between system SP and orifice tube SP

*PERMISSION UNITED SHEET METAL

23.8 FUME HOODS
.1 Overview

This standard of testing, adjusting, and balancing of Fume Hood System is a demonstration of the Associated Air Balance Council's dedication to the safety and welfare of the public.

The term Fume Hoods as used in this specification refers to the enclosed cabinets that are used in laboratories to contain experiments involving various toxic or radioactive levels. The ventilation systems that serve the Fume Hoods are designed to remove the toxic products of the experiments safely, and to prevent the flow of unwanted or dangerous fumes from flowing into the room. These criteria must be met while the work is being done by the operator on the apparatus within the hood enclosure, and with the hood door partially open.

The importance of proper performance of Fume Hood Systems is obvious. The threat of contamination of the operating personnel and laboratories with toxic products is always possible with improper performance. In addition to the health hazards, there is frequently the danger of explosions. There are other factors that must be considered in establishing the desired performance of these systems. These include the amount of fresh air that must be introduced into the building in order to balance the Fume Hood system. The cost of circulating the large quantities of fresh air is a considerable item in the overall cost of laboratory facilities. The thrust of modern design of Fume Hood Systems is to provide the environmentally secure conditions for a facility to operate at the most economical cost.

The importance of the Test and Balance Agency in implementing the Engineer's design cannot be overstressed. The adjustment of total airflow quantities, both exhaust and supply to the hoods, as well as the face velocities and spillage of fumes at the Fume Hood openings are crucial to the effective performance of the Fume Hood System.

It is important to note that while there are two types of Fume Hoods in use today, the type of work being done within the hoods can vary widely. The toxic levels generated within the hood can range from a low level that may be odorous rather than dangerous, to low level radioactive materials or to high chemical toxicity levels. The toxic levels of the work to be done in the Fume Hoods will determine the type and design of the ventilation system that will be serving the facility.

They will also determine the extent of the Test and Balance procedure that will be required.

This section includes a discussion of the types of Fume Hood Systems that are in use and specifications for the Test and Balance of the Fume Hood Systems for different types of toxic substances that are in use in the industry today.

.2 General Discussion
A. Types of Fume Hoods

There are two types of Fume Hoods in use at this time. One introduces a source of makeup or ventilation air directly to the Fume Hood and uses a minimum amount, if any, of the conditioned air from the laboratory space. The second type draws all of its air into the hood from the laboratory space.

The considerations that affect the type of system that is utilized include the toxicity of the processes to be used, the initial cost of the HVAC installation, and the cost of operating the HVAC system. Fume Hood Systems that do not provide self-contained, make up air directly to the Fume Hoods must have sufficient outside air introduced to the space

---

AABC HOOD CERTIFICATION

Test and Balance Agency _____

Hood No. _____ Date _____

Size _____ Average Face Velocity _____

Lowest Velocity Reading _____

CFM _____ TBE _____

Instrument _____

Date Calibrated _____

Fig. 23.8.1: Typical hood certification certificate

through the HVAC system. This type of installation adds additional tonnage to the refrigeration equipment for cooling and to the heating capacity. In systems that provide make-up air directly to the Fume Hoods, the makeup air may not have to be completely conditioned to satisfy the ventilation requirements of the Fume Hood System.

In both types of Fume Hood Systems the safety of personnel and protection of life and property are exactly the same. Because of these ongoing requirements, hood operating tests should be performed at least once a year. Inspection and test result certificates (Fig. 23.8.1) should be placed in a conspicuous location by the AABC Test and Balance Agency.

B. Toxicity Levels

In general, the toxicity level of the work done within the hood is used to determine the air velocity across the face of the hood opening, and therefore the total amount of exhaust air per fume hood. Following is a recommended schedule of face velocities for different toxicity levels:

Low toxic levels	50 FPM
Average toxic levels	75 FPM
Low level radioactive materials	100 FPM
Medium or High Chemically toxic levels	150 FPM

Generally 100 FPM is satisfactory for most applications and provides sufficient airflow to satisfy most laboratory conditions. The Engineer will determine the exhaust airflow rates that will provide the required health and personnel protection with the most economical design.

In addition to the exhaust air quantities, attention must be given to the placement of air conditioning outlets in relation to the fume hoods. Outlets close to the hoods will cause excessive draft conditions and in some cases short circuiting of conditioned air. Optimum application is to place air outlets so that the hoods draw air across the room without creating drafts. When the hoods have self contained air supply systems, attention to the details described above may not be needed.

C. Description of Tests to be Performed by the Test and Balance Agency

The scope of the tests to be performed by the Test and Balance Agency shall be determined by the Engineer and shall be in accordance with the toxicity of the work to be done in the Fume Hoods and the safety standards that must be established for each facility.

There are four separate tests that make up a complete test and balance procedure for Fume Hoods. Following is a detailed specification for each of the tests.
1. Determine and adjust the exhaust airflow.
2. Determine and adjust face velocity, across the hood face or opening.
3. Determine and adjust make-up air conditions.
4. Determine and adjust spillage conditions at hood opening.

.3 Specifications for the Balancing Procedure to be Performed by the AABC Agency

The Test and Balance Agency must perform these four tests to provide a satisfactory balanced condition.
A. Exhaust Airflow
   1. The fume hood face plate or door shall be set in the position normally used by the operating personnel.

23.12
Chapter 23—Special Systems

2. The exhaust CFM shall be measured in the exhaust duct connection to the exhaust fan. This shall be accomplished by measuring airflow using the pitot traverse method. For velocities under 1000 FPM the Micromanometer shall be used. For velocities over 1000 FPM the inclined gage, or the magnehelic gage and pitot tube shall be used. Figure 23.8.2 shows proper positions to determine exhaust fan readings.

3. Any holes drilled for testing shall be sealed as specified by the Engineer.

B. Hood Face Velocity
1. Upon completion of setting exhaust total air, the AABC Test and Balance Agency shall test the face velocity across the hood opening to determine correct flow velocities (Fig. 23.8.3). Use of an electrically operated anemometer is recommended for this test. The instrument used shall be accurate to a minimum velocity of 25

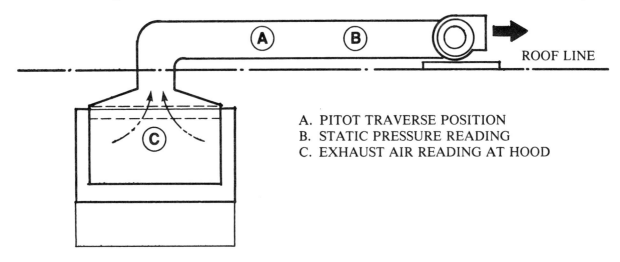

A. PITOT TRAVERSE POSITION
B. STATIC PRESSURE READING
C. EXHAUST AIR READING AT HOOD

Fig. 23.8.2: Positions for exhaust duct readings

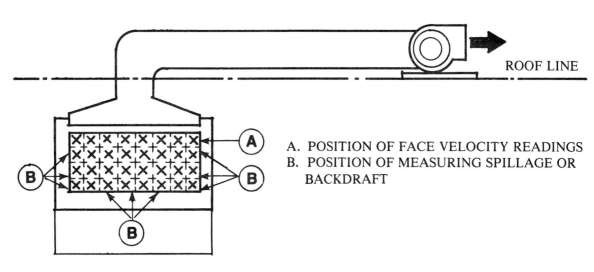

A. POSITION OF FACE VELOCITY READINGS
B. POSITION OF MEASURING SPILLAGE OR BACKDRAFT

Fig. 23.8.3: Positions for face velocities; spillage, and backdraft

23.13
SECTION IV—SPECIFICATIONS

FPM. Tests shall be made with no explosive atmosphere present and with the hood air supply system in operation, if it is part of installation.

2. Hood baffles, shall be adjusted during face velocity tests, so that maximum required face velocities are obtained.

C. Make-up Air

1. Turn on the make-up air supply fan and adjust to a percentage of the exhaust air volume measured at exhaust fan (normally 70%). The readings for make-up air shall be determined by taking static pressure readings across a calibrated orifice in the auxiliary air duct, or by calculating the airflow using the average duct velocity as obtained from a pitot tube traverse.

2. Repeat the test of the hood average face velocity. Adjust the hood baffles, fan drives, and other parts of the system as necessary to provide the specified average face velocity and the specified auxiliary air supply percentage. Repeat tests and adjust the system until fume hood performance and make-up air are in compliance with the specifications.

D. Spillage and Backdraft

1. Make a complete traverse of the hood face with a cotton swab dipped in titanium tetrachloride to determine that a positive flow of air is entering the hood over the entire hood face. No reverse flows or dead air spaces shall be permitted.
CAUTION: Do not allow the titanium tetrachloride to get on the skin or mix with water. When mixed with water, titanium tetrachloride turns into hydrochloric acid.

2. Paint a strip of titanium tetrachloride along each end and across the working surface of the hood, in a line parallel with the hood face and 6" back into the hood from the hood face to determine that no reverse flows of air exist. The flow of smoke shall be directly to the rear of the hood without swirling turbulence or reverse flows.

3. A smoke bomb shall be discharged within the hood area to observe the exhaust capacity of the hood and its design efficiency. The jet of smoke shall show no reverse flows of air when discharged across the bottom of the hood or at the ends of the hoods. The smoke shall be overcome and carried to the rear of the hood and out of the hood even when the jet of smoke is discharged directly toward the front of the hood. Place the smoke bomb on the hood working surface and close the sash (hood door) to determine that sufficient air is passing through the working area of the hood to dilute and exhaust the smoke out of the hood.

4. Place a pan of hot water in the hood at the approximate center of the working area and add dry ice to make heavy white fumes. The fumes shall be carried directly to the rear of the hood. No reverse flows of fumes along the work surfaces shall be permitted.

5. Adjust the sash (hood door) to a position 3" above the bottom of the hood and measure the velocity of the air entering the hood. The velocity shall not be more than $4^{1}/_{2}$ times nor less than 4 times the specified average face velocity.

6. For hoods with auxiliary air, with the sash in the open position discharge a smoke bomb into the auxiliary air supply duct ahead of the blower, to ensure that the smoke is thoroughly mixed with the auxiliary air. Observe the flow of smoke and air down and into the hood front to verify that at least 85% of the auxiliary air is captured. None of the smoke and air shall escape into the room.
7. With sash in the closed position, discharge a smoke bomb into the auxiliary air duct as in item 6. Observe that 100% of the smoke shall be captured and drawn directly into the hood through the bypass. None of the smoke and auxiliary air supply shall be projected down the front of the hood.
8. Check sash hood door operation by raising and lowering sash. Use one hand and grip the sash at the extreme right end. Repeat at the extreme left end of sash. Sash shall glide smoothly and freely and hold at any height without creeping, assuring proper counterbalance. No metal-to-metal contact shall be allowed.

E. Special Tests
1. Perchloric Exhaust System Washdown
   Conduct test of exhaust system washdown of Perchloric Fume Hoods, fans and associated ductwork. Activate the system including hood, duct, fan and vertical discharge duct. Record test and note any discrepancies.
2. Special Filter Systems
   Special filter systems for hoods shall be tested for proper operation and adjusted to maintain constant air volume through the system regardless of filter resistance. Flow shall be tested at not less than five different resistances representing clean, 25%, 50%, 75%, and 100% dirt loadings. Variations in airflow shall not be sufficient to adversely effect hood performance.
3. Interlocks
   Test all fume hood systems in conjunction with associated systems such as auxiliary air, supply air and exhaust air as required to determine that the electrical interlocks specified have been provided and are operating correctly.

.4 Submittals

A. The Report shall contain the following general data in a format selected by the AABC Test and Balance Agency.
   1. Project No.
   2. Contract No.
   3. Project Title
   4. Project Location
   5. Project Architect: (Firm name and address)
   6. Project Mechanical Engineer: (Firm name and address)
   7. Test and Balance Engineer: (Name)
   8. Testing Diagnosis and Analysis by: (Name)
   9. AABC Test and Balance Agency: (Firm name and address)
   10. Fume Hood Manufacturer: (Name and address)
   11. Mechanical Subcontractor: (Name and address)
   12. General Contractor: (Name and address)
   13. Dates that tests were performed and date of Report
   14. Certification

B. The Report shall contain the following specific data on each fume hood in format selected by AABC Testing and Balance Agency.
   1. Project No.
   2. Date

SECTION IV—SPECIFICATIONS

3. Manufacturer's name, model number, and serial number.
4. Individual reference of contract identification number, location, and specified design face velocity.
5. Exhaust Fan and Motor
   a. Fan Manufacturer, model, serial number
   b. Fan description: Such as fan type, size, arrangement, class, discharge, type sheave and drive, speed (RPM), specified design total exhaust CFM (high and low if two speed)
   c. Motor Manufacturer, model, serial number
   d. Motor description: Such as HP, voltage, phase, cycles, rated amperes, running amperes, speed (RPM).
6. Make-up Air Supply Fan and Motor
   a. Fan Manufacturer, model, serial number
   b. Fan description: Such as fan type, size, arrangement, class, discharge, type sheave and drive, speed (RPM), specified design total exhaust CFM (high and low if two speed)
   c. Motor Manufacturer, model, serial number
   d. Motor description: Such as HP, voltage, phase, cycles, rated amperes, running amperes, speed (RPM).
7. Test Conditions of Hood
   a. Width and depth of hood working surface
   b. Height and width of hood door (sash) opening
   c. Room temperature (degrees F)
   d. Auxiliary air supply temperature (degrees F)
   e. Size and distance of HVAC outlets and returns which might affect hood performance if operating
   f. HVAC operating or not operating during test
   g. Number and location of hoods in use at time of test
   h. Position (open or closed) of room doors
   i. Equipment in hood (if any) when readings taken
8. Performance Test Data (adjustments completed)
   a. Position of fume door (normally full open)
   b. Exhaust volume rate (CFM) measured. Include average duct velocity and cross sectional area of duct used in calculations.
   c. Exhaust volume rate (CFM) measured at hood duct opening. Include average duct velocity and cross sectional area of duct used for calculations.
   d. Sketch of hood door opening in 4 inch increments with each face velocity reading.
   e. Average face velocity. Compare with specified design face velocity.
   f. Exhaust volume rate (CFM) calculated from face velocity measurements. Compare with exhaust volumes of 8.b and c.
   g. Make-up air supply volume rate (CFM). Include average duct velocity and cross sectional area of duct used for calculations.
   h. Make-up air supply rate as a percentage of exhaust volume rate.
   i. Whether reverse flows or dead air spaces were observed at hood face. (Titanium Tetrachloride Test).
   j. Whether reverse flows were observed at each end of the working surface and across the working surface of hood. (Titanium Tetrachloride Test).

Chapter 23—Special Systems

k. Observations and results of hood smoke test with hood door open and door closed.
l. Observations and results of hood dry-ice test.
m. Average face velocity with hood door open 3 inches. Compare with specified limitations.
n. Observations and results of auxiliary air supply smoke test with hood door open and closed.
o. Result of test of exhaust duct airflow.
p. Results of exhaust system wash down (Perchloric System).
q. Brief Summary of tests.

Single sheets of specifications suitable for photocopying are available from AABC National Headquarters, upon request.

CHAPTER 24

TEMPERATURE CONTROL SYSTEMS

24.1 OVERVIEW

Automatic temperature controls are so intimately involved in the process of **Total System Balance,** that the responsibility of the AABC Test and Balance Agency must be defined. This chapter specifies the responsibility of the AABC Test and Balance Agency as related to temperature control systems.

The responsibility of the AABC Test and Balance Agency is to verify control system operation as specified, and report on any installation problems observed. The AABC Test and Balance Agency shall limit its activities to setting controls to a proper fixed mode to prevent any changes during the balancing procedure. This also provides a verification of control operation which is valuable to all parties. Physical changes in the control system—such as relocating sensors, or calibrating controllers—is the responsibility of the control contractor. The AABC Test and Balance Agency shall work closely with the control contractor to identify and correct problems.

24.2 SPECIFICATIONS

In the process of **Total System Balance,** The AABC Test and Balance Agency shall:

.1 Work with the temperature control contractor to ensure the most effective total system operation within the design limitations, and to obtain mutual understanding of intended control performance.

.2 Verify that all control devices are properly connected.

.3 Verify that all dampers, valves, and other controlled devices are operated by the intended controller.

.4 Verify that all dampers and valves are in the position indicated by the controller (open, closed, or modulating).

.5 Verify the integrity of valves and dampers in terms of tightness of close-off and of full-open position. This includes dampers in multizone units, mixing boxes and VAV terminals.

.6 Check that all valves are properly installed in the piping system in relation to direction of flow and location.

.7 Check the calibration of all controllers.

.8 Verify the proper application of all normally open and normally closed valves.

.9 Check the locations of all thermostats and humidistats for potential erratic operation from outside influences such as sunlight, drafts, or cold walls.

.10 Check the locations of all sensors to determine whether their position will allow them to sense only the intended temperatures or pressures of the media.

.11 Check that the sequence of operation for any control mode is in accordance with approved shop drawings. Verify that no simultaneous heating and cooling occurs. Observe that heating cannot take place at VAV reheat terminals until the unit is at minimum CFM.

.12 Verify that all controller set points meet the design intent.

.13 Check all dampers for free travel.

.14 Verify the operation of all interlock systems.

.15 Perform all system verification to assure the safety of the system and its components.

CHAPTER 25

PRECONSTRUCTION PLAN CHECK AND CONSTRUCTION REVIEW

25.1 OVERVIEW

This chapter covers the minimum procedures for the Preconstruction Plan Check before a project is begun, and for Construction Review during the project.

The concept of Preconstruction Plan Check and Construction Review as an essential part of **Total System Balance** was developed by Associated Air Balance Council. This concept is part of the AABC commitment to assure the client of professional, quality work.

A Preconstruction Review, plus inspection during construction by the Test and Balance Agency can contribute substantially to a system that will perform more efficiently and more economically. These review procedures will not only assist in achieving the design intent; they will also simplify the **Total System Balance** procedures and reduce the time required for the project.

25.2 PRECONSTRUCTION PLAN CHECK

The Test and Balance Agency is the most qualified to determine if **Total System Balance** can be accomplished within the design. The Preconstruction Plan Check allows the Test and Balance Agency to assist the designer in achieving the design intent.

The Preconstruction Plan Check which includes a Preliminary Review occurs as soon as possible after the Test and Balance Agency receives the responsibility for the project. If the review is for new construction, it must occur before actual work is started on the project. If this is not done, required modifications will be costly and will result in delays. If the review is for an existing system, the actual system must be physically inspected and analyzed in addition to the procedures indicated in this section.

.1 Purpose

The purpose of the Preliminary Review is to:
A. Review the project documents and assume that the design intent is completely understood by the Test and Balance Agency.
B. Identify potential problems from the viewpoint of **Total System Balance.**
C. Allow the **Total System Balance** Agency to apply its specialized knowledge and experience. This will improve the **Total System Balance** and achieve the most effective performance in accordance with the design intent.
D. Develop a written Preliminary Report which lists the changes and how they should be made to allow the most effective **Total System Balance.**

.2 Items to Review

The following documents should be reviewed:
A. Contract Documents
   1. Drawings
   2. Specifications
   3. Addenda
B. Submittal Data
C. Shop Drawings
D. Temperature Control Drawings

.3 Review of Specifications

Specifications should be reviewed for:
A. Scope of work.
B. Special requirements.
C. Items that will make balancing difficult or impossible.

Those sections of the specifications that are specific to **Total System Balance** should be reviewed in detail. In addition, all other sections of the specifications should be reviewed for their possible effect on **Total System Balance**. Typical items are equipment specifications, procedures of construction, and scheduling.

.4 Review of Drawings

Drawings should be reviewed for:
A. Potential problems for **Total System Balance.** (Balancing devices, general system layout, architectural features.)
B. The most effective **Total System Balance** procedures.
C. Potential problems that may affect future operation.
D. Scheduling and coordination needs with other trades.

.5 Review of Submittal Data

Submittal data should be reviewed for completeness and for conformity to the contract documents as well as for potential problems for **Total System Balance.** Typical submittal data items are:
A. Equipment manufacturers' submittals.
B. Special instructions for use of balancing devices.
C. Pump curves.
D. Factors for flow meters.
E. Limitations affecting accuracy of equipment (such as length of straight pipe for flow meters).

.6 Review of Shop Drawings

Contractor's shop drawings should be reviewed for potential problems, and especially to see whether these drawings include everything required in the specifications but not shown on the contract drawings. Careful checks should be made for conformity to "blanket" statements in the specifications such as, ". . . all branch or zone ducts shall have a volume damper in close proximity to the main trunk duct."

.7 Review of Temperature Control Drawings

Temperature Control Drawings should be reviewed for:
A. Understanding the system.
B. Determining the most effective **Total System Balance** procedure for minimum control manipulation, and to avoid disturbing the calibration of control devices.
C. Determining what manipulation of controls will be necessary for **Total System Balance,** and whether the control contractor should be consulted in this regard.

25.3 CONSTRUCTION REVIEW

Construction Reviews consisting of on-site visits **must be made during progress of the project.** The frequency of the visits shall be specified.

Construction Reviews assure that balancing devices such as volume dampers, flow metering devices, pressure measuring and temperature measuring stations, and similar devices are being installed as required for **Total System Balance** as well as for future operation of the system.

.1 Purpose

The purpose of Construction Reviews is to:
A. Identify potential problems for performing **Total System Balance.**
B. Identify modifications which will aid **Total System Balance.**

C. Schedule and coordinate **Total System Balance** with other work.
D. Identify any conditions that could create a hazardous environment for building occupants.

.2 Typical Activities
A. Check for necessary balancing hardware in place (dampers, flow meters, valves, pressure taps, thermometer wells, etc.)
 1. Located properly and accessible.
 2. Installed correctly.
B. Identify and evaluate any variations from system design.
C. Gather data from equipment nameplates. (Usually easier to read at this time. Also may identify problems.)
D. Identify and report possible restrictions in systems (closed fire dampers, long runs of flexible duct, poorly designed duct fittings, etc.)
E. Verify that construction progress will not delay **Total System Balance**.
F. Identify the best location for duct pitot tube traverses.
G. Identify scaffolding needs.

The specifications in this chapter are interrelated with those in other chapters of this section and must be properly combined to form a complete set of specifications. See Chapter 16 for a checklist of elements to form a complete set of specifications.

CHAPTER 26

REPORTS AND REPORT FORMS

26.1 OVERVIEW

Material contained in this chapter defines the minimum data required for the **Total System Balance Report**.

Specific information regarding the report format, sections of the report, report outline, and recommended report forms are also included. The test results contained in these sections are vital to a comprehensive and complete **Total System Balance Report**. The report is a complete record of design, preliminary measurement, and final test data. The report shall also indicate any discrepancies between the project specifications and the test data. It shall also include reasons for any discrepancies.

26.2 GENERAL

The report data outlined in this chapter is the minimum information required for a complete Test and Balance Report. Additional information shall also be included as specified in the contract documents, and as applicable to the project.

26.3 DEFINITION OF A REPORT

A report is a record of actual test and balance results. This report shall reflect actual, tested, and observed conditions of all systems and components during **Total System Balance**. The report shall be certified, dated, and signed by the certified Test and Balance Engineer.

A report also contains design data as well as the normal operating conditions of the building systems that are specified for **Total System Balance**. It also provides sufficient data pertaining to the instruments that were used and the system operating modes during balancing to allow any reported data to be repeated.

A **Total System Balance** project shall not be considered complete until the owner's representative has been provided with a final Test and Balance Report that thoroughly describes the operation of the systems.

26.4 REPORT FORMAT

Copies of the report shall be submitted in the quantity specified. Reports shall be neatly typed and bound. Each data sheet in the report shall contain the AABC Logo and shall be identified by date, page number, system designation, system location, and project name.

If specified, this report shall include a statement of temperature control verification as applicable to the Test and Balance procedure in Accordance with the AABC National Standards, 1982 (Chap. 24).

26.5 SECTIONS OF REPORT

The Report shall contain the following sections as applicable to the project specifications:

- **General Information and Summary**
- **Air Systems**
- **Hydronic Systems**
- **Temperature Control Systems**
- **Special Systems**
- **Sound and Vibration Systems**

26.6 REPORTS

The Report shall contain, as a minimum, all of the information included in the following recommended forms.

Chapter 26—Reports and Report Forms

TEST FORM INDEX

TITLE	FORM NUMBER	PAGE NUMBER
Cover Sheet	82010	26.3
Instrument List	82020	26.4
Air Moving Equipment		
Air Moving Equipment Test Sheet	82030	26.5
Exhaust Fan Data Sheet	82031	26.6
Static Pressure Profile	82032	26.7
Return Air/Outside Air Data	82033	26.8
Fan & Motor Pulley	82034	26.9
Duct Traverse Readings	82035	26.10
Duct Traverse Zone Totals	82036	26.11
Air Monitoring Station Data	82037	26.12
Air Terminal		
Air Distribution Test Sheet	82040	26.13
Terminal Units	82041	26.14
Induction Units	82042	26.15
Electric Duct Heater	82050	26.16
Pump Data Sheet	82060	26.17
Cooling Tower	82070	26.18
Heat Transfer Equipment		
Chillers	82080	26.19
Air Cooled Condenser	82081	26.20
Primary Heat Exchanger	82090	26.21
Heat Transfer Elements		
Cooling Coil Data	82100	26.22
Heating Coil Data	82101	26.23
Flow Measuring Station	82102	26.24
Sound		
Sound Level Report	82300	26.25
Octave Band Chart	82301	26.26
Vibration Test Data		
Air Handling Unit	82400	26.27
Centrifugal Fan	82401	26.28
Horizontal Split Case Pump	82402	26.29
In Line Fan	82403	26.30
Utility Fan	82404	26.31
End Suction Pump	82405	26.32
Vaneaxial Fan	82406	26.33
Duct Leak Test	82500	26.34
Combustion Test	82600	26.35

26.3
SECTION V—PROCEDURES

AABC LOGO

COMPANY NAME

COMPANY   COMPANY
ADDRESS   COMPANY LOGO   TELEPHONE #

TEST AND BALANCE REPORT

Project _____

Location _____

Architect _____

Engineer _____

Contractor _____

Project Number _____

This is to certify that (COMPANY) has balanced the systems described herein to their optimum performance capabilities. The testing and balancing has been performed in accordance with the standard requirements and procedures of the Associated Air Balance Council and the results of these tests are herein recorded.

Associated Air Balance Council Certification Number _____

Date _____   _____
                                                                              Test & Balance Engineer

orm #82010

AABC LOGO

Date _____

Page _____ of _____

Project Name _____

INSTRUMENT LIST

INSTRUMENT	MANUFACTURER	MODEL	SERIAL NUMBER	RANGE	CALIBRATION DATE
1.					
2.					
3.					
4.					
5.					
6.					
7.					
8.					
9.					
10.					
11.					
12.					
13.					
14.					
15.					
16.					
17.					
18.					
19.					
20.					

Remarks _____

_____

Form #82020

26.5
SECTION V—PROCEDURES

AABC LOGO

Date _____

Page _____ of _____

Project Name _____

AIR MOVING EQUIPMENT TEST SHEET

SYSTEM	
Equipment Location	
Area Served	
Equipment Manufacturer	
Model	
Serial Number	

	Specified	Actual		Specified	Actual
Total CFM – Fan					
Total CFM – Outlet					
R/A CFM					
O/A CFM					
Total Static Pressure (Total/External)					
Inlet Pressure					
Discharge Pressure					
Fan RPM					

	Specified	Actual		Specified	Actual
Motor Manufacturer					
Motor HP/BHP					
Phase					
Voltage					
Amperage					
Motor RPM					
Motor Service Factor					
Starter Heater Elements					

Motor Sheave & No. Grooves		
Fan Sheave & No. Grooves		
Belts		

Remarks _____
_____

Form #82030

AABC LOGO

Date _____

Page _____ of _____

Project Name _____

EXHAUST FAN DATA SHEET

SYSTEM		
Equipment Location		
Area Served		
Equipment Manufacturer		
Model		
Serial Number		

	Specified	Actual	Specified	Actual
Total CFM – Fan				
Total CFM – Outlet				
Total Static Pressure* (Total/External)				
Inlet Pressure				
Discharge Pressure				
Fan RPM				

	Specified	Actual	Specified	Actual
Motor Manufacturer				
Motor HP / BHP				
Phase				
Voltage				
Amperage				
Motor RPM				
Motor Service Factor				
Starter Heater Elements				

Motor Sheave & No. Grooves		
Fan Sheave & No. Grooves		
Belts		

*Not always required or applicable.

Remarks _____

_____

_____

Form #82031

26.7
SECTION V—PROCEDURES

AABC LOGO

Date _____

Page _____ of _____

Project Name _____

STATIC PRESSURE PROFILE

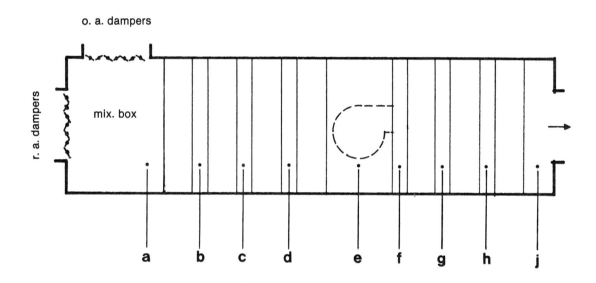

UNIT	a	b	c	d	e	f	g	h	j

Remarks _____

Form #82032

AABC LOGO

Date _____

Page _____ of _____

Project Name _____

RETURN AIR/OUTSIDE AIR DATA

Unit Number			
Design CFM			
Actual CFM			
Design R.A.			
Actual R.A. CFM			
Design O.S.A.			
Actual O.A. CFM			
R.A. Temperature			
O.S.A. Temperature			
Required M.A. Temperature			
Actual M.A. Temperature			
Design O.S.A./R.A. Ratio			
Actual O.S.A./R.A. Ratio			

Remarks _____

_____

_____

_____

_____

Form #82033

SECTION V—PROCEDURES　26.9

AABC LOGO

Date _____

Page _____ of _____

Project _____

FAN & MOTOR PULLEY
(For Field Use Only)

C.C.

Unit No. _____

Fan Sheave:

　Diameter _____

　RPM _____

Belts:

　Size _____

　Quantity _____

Motor Sheave:

　Diameter _____

　RPM _____

Required:

　RPM _____

　Sheave _____

　Belts _____

Center to Center:

　Maximum _____

　Minimum _____

　Actual _____

Remarks _____
_____

Form #82034

AABC LOGO

Date _____

Page _____ of _____

Project _____

System No. _____ Traverse Location _____

Area Served _____

DUCT TRAVERSE READINGS

POINT NO.										
A										
B										
C										
D										
E										
F										
G										
H										
I										
J										
SUB TOTAL										

TOTAL FPM ÷ NO. READINGS = AVERAGE FPM x AREA = TOTAL CFM

_____ ÷ _____ = _____ x _____ = _____

	1	2	3	4	5	6
A	+	+	+	+	+	+
B	+	+	+	+	+	+
C	+	+	+	+	+	+
D	+	+	+	+	+	+

READING POINTS FOR PITOT MEASUREMENTS

Remarks _____

_____

_____

_____

DESIGN	DUCT SIZE	
	AREA	
	FPM	
	CFM	

TEST DATA	DUCT SIZE O.D.	
	DUCT SIZE I.D.	
	AREA	
	CENTER LINE S.P.	
	AVERAGE FPM	
	CFM MEASURED	
	TEMP CORRECTION	
	ALT CORRECTION	
	CFM STD. COND.	

FORM #82035

26.11
SECTION V—PROCEDURES

AABC LOGO

Date _____

Page _____ of _____

Project _____

DUCT TRAVERSE ZONE TOTALS

System Zone/Branch	Duct Size	Area Sq. Ft.	Design		Test			
			FPM	CFM	Test 1 FPM	Test 2 FPM	CFM	Static Pressure

Remarks _____

_____

Form #82036

AABC LOGO

Date _____

Page _____ of _____

Project _____

AIR MONITORING STATION DATA

Station Number	System	Area Served	Size	Area Sq. Ft.	Design		Test		
					FPM	CFM	Test 1 FPM	Test 2 FPM	CFM

Remarks: _____

_____

Form #82037

26.13
SECTION V—PROCEDURES

ABC LOGO

Date _____

Page _____ of _____

Project _____

System _____ Floor # _____

AIR DISTRIBUTION TEST SHEET

Terminal Number	Room Number	Terminal		Factor	Design		Test—FPM or CFM			Final	
		Type	Size		FPM	CFM	Test 1	Test 2	Test 3	FPM	CFM

Remarks _____
_____

Form #82040

AABC LOGO

Date _____

Page _____ of _____

Project _____

TERMINAL UNITS

System _____

Manufacturer _____

| CONSTANT | | VARIABLE | | SINGLE | | DUAL | |

Box Number	Location	Model	Size	Min. S.P.	Min. CFM Design	Max. CFM Design	Min. CFM Actual	Max. CFM Actual

Remarks: _____

_____

Form #82041

26.15
SECTION V—PROCEDURES

AABC LOGO

Date _____

Page _____ of _____

Project _____

System _____

INDUCTION UNITS

Unit	Location	Model	Size	Design CFM	Design Nozzle Pr. W.G.	Nozzle Pressure Readings		Final Nozzle Pr. W.G.	Final CFM
						No. 1	No. 2		

Remarks _____

Form #82042

AABC LOGO

Date _____

Page _____ of _____

Project _____

ELECTRIC DUCT HEATER

COIL NO.	LOCATION	MODEL NO.	DESIGN DATA				TEST DATA					
			KW	STAGES	VOLTS PHASE	AMPS	VOLTAGE			AMPERAGE		
							T1-T2	T2-T3	T3-T1	T1	T2	T3

Remarks _____

Form #82050

26.17
SECTION V—PROCEDURES

AABC LOGO

Date _____

Page _____ of _____

Project _____

PUMP DATA SHEET

PUMP NO.	
MANUFACTURER	
SIZE	
IMPELLER	
SERVICE	

TEST DATA	GPM	FT. HD.	BHP
DESIGN			
ACTUAL			
DISCHARGE			
SUCTION			
△p	x 2.31 =		FT. HD.

BLOCK OFF			
DISCHARGE			
SUCTION			
△p	x 2.31 =		FT. HD.

MOTOR MFG.			
FRAME			
H.P.			
RPM			
AMPS	ACT:	/	/
VOLTS	ACT:	/	/
REMARKS:			

PUMP NO.	
MANUFACTURER	
SIZE	
IMPELLER	
SERVICE	

TEST DATA	GPM	FT. HD.	BHP
DESIGN			
ACTUAL			
DISCHARGE			
SUCTION			
△p	x 2.31 =		FT. HD.

BLOCK OFF			
DISCHARGE			
SUCTION			
△p	x 2.31 =		FT. HD.

MOTOR MFG.			
FRAME			
H.P.			
RPM			
AMPS	ACT:	/	/
VOLTS	ACT:	/	/
REMARKS:			

AABC LOGO

Date _____

Project _____

Page _____ of _____

COOLING TOWER

Tower # _____

Make _____

Model # _____

Serial # _____

Rated Capacity _____

Motor Manufacturer _____

DATA	Specified	Actual	Specified	Actual	Specified	Actual
HP						
Volts						
RPM						
Amps						
S.F. Factor						
Over Load Elements						
Air WB: On _____	°F	°F	°F	°F	°F	°F
Off _____	°F	°F	°F	°F	°F	°F
Air DB (amb.) _____	°F	°F	°F	°F	°F	°F
Cond. Water On _____	°F	°F	°F	°F	°F	°F
Off _____	°F	°F	°F	°F	°F	°F
GPM						
Fan RPM						
Drive: Motor Sheave						
Fan Sheave						
Belt						
Gear Manufacturer						
Model						

Remarks _____

Form #82070

SECTION V—PROCEDURES

AABC LOGO

Date _____

Page _____ of _____

Project _____

CHILLERS

UNIT NO.		
MANUFACTURER		
CAPACITY		
MODEL	SERIAL NO.	

COOLER	DESIGN	ACTUAL
ENT. WATER TEMP.		
LVG. WATER TEMP.		
PRESS. DROP, FT.		
GPM		

CONDENSER	DESIGN	ACTUAL
ENT. WATER TEMP.		
LVG. WATER TEMP.		
PRESS. DROP, FT.		
GPM		

ELECTRICAL	DESIGN	ACTUAL
VOLTAGE—T1—T2		
T2—T3		
T3—T1		
AMPERAGE T1		
T2		
T3		
REMARKS:		

UNIT NO.		
MANUFACTURER		
CAPACITY		
MODEL	SERIAL NO.	

COOLER	DESIGN	ACTUAL
ENT. WATER TEMP.		
LVG. WATER TEMP.		
PRESS. DROP, FT.		
GPM		

CONDENSER	DESIGN	ACTUAL
ENT. WATER TEMP.		
LVG. WATER TEMP.		
PRESS. DROP, FT.		
GPM		

ELECTRICAL	DESIGN	ACTUAL
VOLTAGE—T1—T2		
T2—T3		
T3—T1		
AMPERAGE T1		
T2		
T3		
REMARKS:		

Form #82080

AABC LOGO

Date _____

Page _____ of _____

Project _____

AIR COOLED CONDENSER

Unit No.					
Location					
Manufacturer					
Model					
Serial					
	DESIGN	ACTUAL		DESIGN	ACTUAL
E.A.T. DB					
L.A.T. DB					

Compressor #								
	DESIGN	ACTUAL	DESIGN	ACTUAL	DESIGN	ACTUAL	DESIGN	ACTUAL
Motor HP								
Amps								
Volts								

Remarks: _____
_____
_____
_____
_____
_____

Form #82081

26.21
SECTION V—PROCEDURES

AABC LOGO

Date _____

Page _____ of _____

Project _____

PRIMARY HEAT EXCHANGER

UNIT DATA	Unit No.	Unit No.	Unit No.	Unit No.
Location				
Service				
Rating, BTU/HR				
Manufacturer				
Model No.				
Serial No.				

	TEST DATA	Design	Actual	Design	Actual	Design	Actual	Design	Actual
STEAM	Press., PSI								
PRIMARY HW	Ent. Temp.								
	Lvg. Temp.								
	Flow, Gpm								
	Press. Drop, Ft.								
SECONDARY HW	Ent. Temp.								
	Lvg. Temp.								
	Flow, Gpm								
	Press. Drop, Ft.								
	Control Setting								

Remarks: _____
_____
_____
_____

Form #82090

AABC LOGO

Date _____

Page _____ of _____

Project _____

COOLING COIL DATA

System								
Location								
Service								
Manufacturer								
	Design	Actual	Design	Actual	Design	Actual	Design	Actual
CFM								
GPM								
Coil P.D., FT.								
E.W.T., °F								
L.W.T., °F								
E.A.T., DB °F								
E.A.T., WB °F								
L.A.T., DB °F								
L.A.T., WB °F								

Remarks: _____

_____

_____

_____

_____

_____

Form #82100

26.23
SECTION V—PROCEDURES

AABC LOGO

Date _____

Page _____ of _____

Project _____

HEATING COIL DATA

System								
Location								
Service								
Manufacturer								
	Design	Actual	Design	Actual	Design	Actual	Design	Actual
CFM								
GPM								
Coil P.D., FT.								
E.W.T., °F								
L.W.T., °F								
E.A.T., DB °F								
L.A.T., DB °F								

Remarks: _____

Form #82101

AABC LOGO

Date _____

Page _____ of _____

Project _____

Manufacturer _____

FLOW MEASURING STATION

Station No.	Location	Size	Model	Design		Test		Final	
				G.P.M.	PD	Test 1 PD	Test 2 PD	PD	G.P.M.

Remarks: _____

_____

Form #82102

26.25
SECTION V—PROCEDURES

AABC LOGO

Date _____

Page _____ of _____

Project _____

SOUND LEVEL REPORT

LOCATION –		OCTAVE BANDS – EQUIPMENT OFF										
	ALL PASS	A	31.5	63	125	250	500	1000	2000	4000	8000	16000
		OCTAVE BANDS – EQUIPMENT ON										
	ALL PASS	A	31.5	63	125	250	500	1000	2000	4000	8000	16000

LOCATION –		OCTAVE BANDS – EQUIPMENT OFF										
	ALL PASS	A	31.5	63	125	250	500	1000	2000	4000	8000	16000
		OCTAVE BANDS – EQUIPMENT ON										
	ALL PASS	A	31.5	63	125	250	500	1000	2000	4000	8000	16000

LOCATION –		OCTAVE BANDS – EQUIPMENT OFF										
	ALL PASS	A	31.5	63	125	250	500	1000	2000	4000	8000	16000
		OCTAVE BANDS – EQUIPMENT ON										
	ALL PASS	A	31.5	63	125	250	500	1000	2000	4000	8000	16000

LOCATION –		OCTAVE BANDS – EQUIPMENT OFF										
	ALL PASS	A	31.5	63	125	250	500	1000	2000	4000	8000	16000
		OCTAVE BANDS – EQUIPMENT ON										
	ALL PASS	A	31.5	63	125	250	500	1000	2000	4000	8000	16000

LOCATION –		OCTAVE BANDS – EQUIPMENT OFF										
	ALL PASS	A	31.5	63	125	250	500	1000	2000	4000	8000	16000
		OCTAVE BANDS – EQUIPMENT ON										
	ALL PASS	A	31.5	63	125	250	500	1000	2000	4000	8000	16000

Remarks: _____

Form #82300

26.26
Chapter 26—Reports and Report Forms

AABC LOGO

Date _____

Page _____ of _____

Project _____

OCTAVE BAND CHART

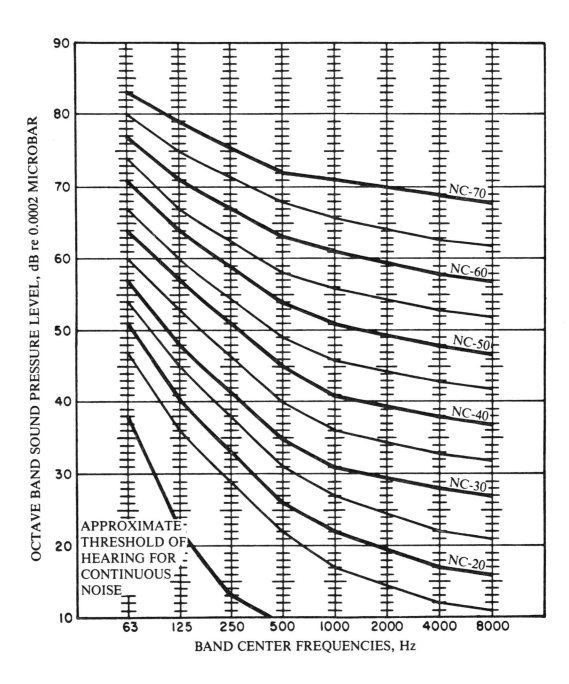

Remarks: _____

Form #82301

26.27
SECTION V—PROCEDURES

AABC LOGO

Date _____

Page _____ of _____

Project _____

VIBRATION TEST DATA
AIR HANDLING UNIT

RPM _____

Motor HP _____

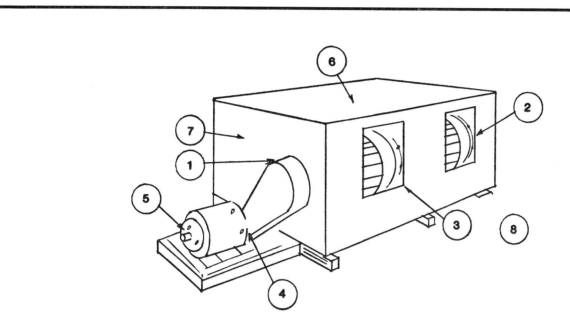

LOCATION OF POINTS		READINGS					
		HORIZONTAL		VERTICAL		AXIAL	
NO.	DESCRIPTION	VEL.	DISP.	VEL.	DISP.	VEL.	DISP.
1.	FAN BEARING, DRIVE END						
2.	FAN BEARING, OPPOSITE END						
3.	MOTOR BEARING, CENTER (IF APPLICABLE)						
4.	MOTOR BEARING, DRIVE END						
5.	MOTOR BEARING, OPPOSITE END						
6.	CASING (BOTTOM OR TOP)						
7.	CASING (SIDE)						
8.	DUCT AFTER FLEXIBLE CONNECTION (DISCHARGE)						
9.	DUCT AFTER FLEXIBLE CONNECTION (SUCTION)						

Normally Acceptable Readings: VEL. _____ ACC. _____

Unusual Conditions At Time Of Test: _____

Vibration Source (If Non Complying): _____

Remarks: _____

Form #82400

26.28
Chapter 26—Reports and Report Forms

AABC LOGO

Date _____

Page _____ of _____

Project _____

VIBRATION TEST DATA
CENTRIFUGAL FAN

RPM _____

Motor HP _____

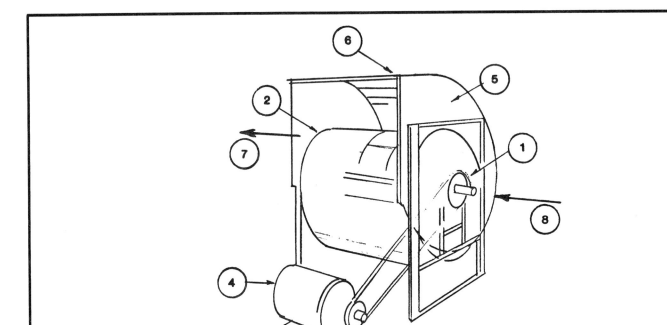

LOCATION OF POINTS		READINGS					
		HORIZONTAL		VERTICAL		AXIAL	
NO.	DESCRIPTION	VEL.	DISP.	VEL.	DISP.	VEL.	DISP.
1.	FAN BEARING, DRIVE END						
2.	FAN BEARING, OPPOSITE END						
3.	MOTOR BEARING, DRIVE END						
4.	MOTOR BEARING, OPPOSITE END						
5.	CASING (TOP)						
6.	CASING (SIDE)						
7.	DUCT OR CASING AFTER FLEXIBLE CONNECTION (DISCHARGE)						
8.	DUCT OR CASING AFTER FLEXIBLE CONNECTION (SUCTION)						

Normally Acceptable Readings: VEL. _____ ACC. _____

Unusual Conditions At Time Of Test: _____

_____

Vibration Source (If Non Complying): _____

_____

Remarks: _____

Form #82401

26.29
SECTION V—PROCEDURES

AABC LOGO

Date _____

Page _____ of _____

Project _____

VIBRATION TEST DATA
HORIZONTAL SPLIT CASE PUMP

RPM _____

Motor HP _____

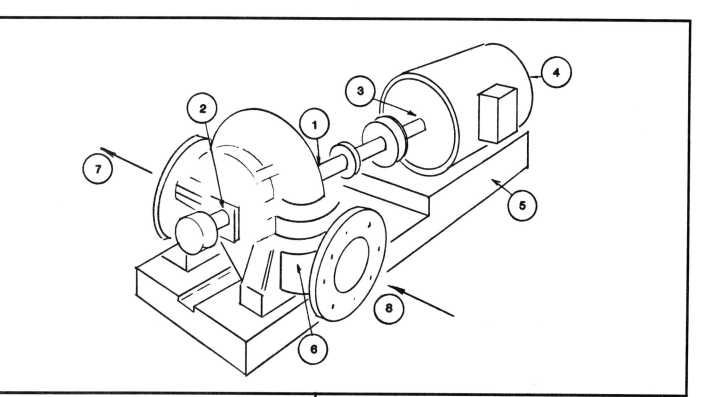

NO.	LOCATION OF POINTS	READINGS					
		HORIZONTAL		VERTICAL		AXIAL	
	DESCRIPTION	VEL.	DISP.	VEL.	DISP.	VEL.	DISP.
1.	PUMP BEARING, DRIVE END						
2.	PUMP BEARING, OPPOSITE END						
3.	MOTOR BEARING, DRIVE END						
4.	MOTOR BEARING, OPPOSITE END						
5.	STRUCTURE (BASE)						
6.	STRUCTURES (CASING)						
7.	PIPE AFTER FLEXIBLE CONNECTION (DISCHARGE)						
8.	PIPE AFTER FLEXIBLE CONNECTION (SUCTION)						

Normally Acceptable Readings: VEL. _____ ACC. _____

Unusual Conditions At Time Of Test: _____

Vibration Source (If Non Complying): _____

Remarks: _____

Form #82402

26.30
Chapter 26—Reports and Report Forms

AABC LOGO

Date _____

Page _____ of _____

Project _____

VIBRATION TEST DATA
IN LINE FAN

RPM _____

Motor HP _____

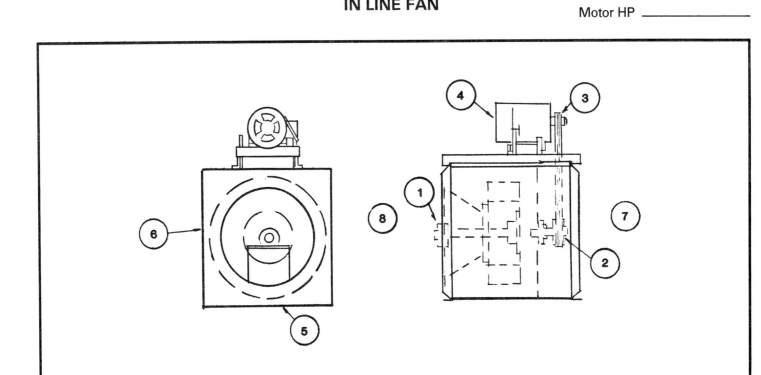

LOCATION OF POINTS		READINGS					
		HORIZONTAL		VERTICAL		AXIAL	
NO.	DESCRIPTION	VEL.	DISP.	VEL.	DISP.	VEL.	DISP.
1.	FAN BEARING, DRIVE END						
2.	FAN BEARING, OPPOSITE END						
3.	MOTOR BEARING, DRIVE END						
4.	MOTOR BEARING, OPPOSITE END						
5.	CASING (BOTTOM OR TOP)						
6.	CASING (SIDE)						
7.	DUCT AFTER FLEXIBLE CONNECTION (DISCHARGE)						
8.	DUCT AFTER FLEXIBLE CONNECTION (SUCTION)						

Normally Acceptable Readings: VEL. _____ ACC. _____

Unusual Conditions At Time Of Test: _____

_____

Vibration Source (If Non Complying): _____

_____

Remarks: _____

Form #82403

26.31
SECTION V—PROCEDURES

AABC LOGO

Date _____

Page _____ of _____

Project _____

VIBRATION TEST DATA
UTILITY FAN

RPM _____

Motor HP _____

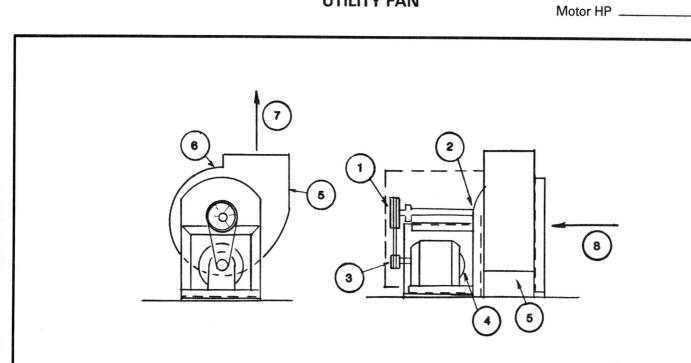

LOCATION OF POINTS		READINGS					
		HORIZONTAL		VERTICAL		AXIAL	
NO.	DESCRIPTION	VEL.	DISP.	VEL.	DISP.	VEL.	DISP.
1.	FAN BEARING, DRIVE END						
2.	FAN BEARING, OPPOSITE END						
3.	MOTOR BEARING, DRIVE END						
4.	MOTOR BEARING, OPPOSITE END						
5.	CASING OR FRAME (DISCHARGE)						
6.	CASING OR FRAME (SUCTION)						
7.	DUCT AFTER FLEXIBLE CONNECTION (DISCHARGE)						
8.	DUCT AFTER FLEXIBLE CONNECTION (SUCTION)						

Normally Acceptable Readings: VEL. _____ ACC. _____

Unusual Conditions At Time Of Test: _____

_____

Vibration Source (If Non Complying): _____

_____

Remarks: _____

Form #82404

26.32
Chapter 26—Reports and Report Forms

AABC LOGO

Date _____

Page _____ of _____

Project _____

VIBRATION TEST DATA
END SUCTION PUMP

RPM _____

Motor HP _____

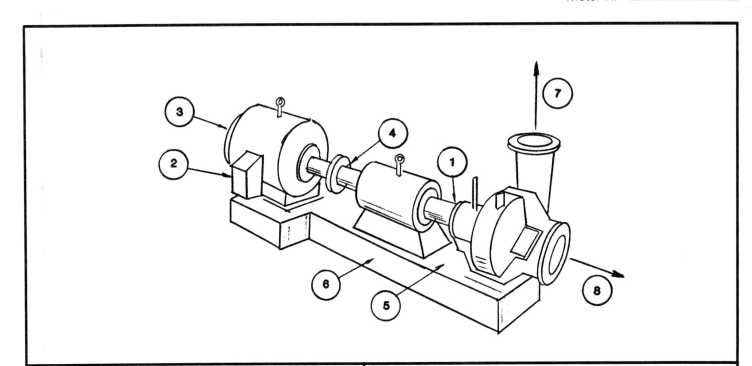

NO.	LOCATION OF POINTS DESCRIPTION	READINGS					
		HORIZONTAL		VERTICAL		AXIAL	
		VEL.	DISP.	VEL.	DISP.	VEL.	DISP.
1.	PUMP BEARING, DRIVE END						
2.	MOTOR BEARING, DRIVE END						
3.	MOTOR BEARING, OPPOSITE END						
4.	COUPLING OR SHAFT SUPPORT						
5.	STRUCTURE (TOP)						
6.	STRUCTURE (SIDE)						
7.	PIPE AFTER FLEXIBLE CONNECTION (DISCHARGE)						
8.	PIPE AFTER FLEXIBLE CONNECTION (SUCTION)						

Normally Acceptable Readings: VEL. _____ ACC. _____

Unusual Conditions At Time Of Test: _____

Vibration Source (If Non Complying): _____

Remarks: _____

Form #82405

26.33
SECTION V—PROCEDURES

AABC LOGO

Date _____

Page _____ of _____

Project _____

VIBRATION TEST DATA
VANEAXIAL FAN

RPM _____

Motor HP _____

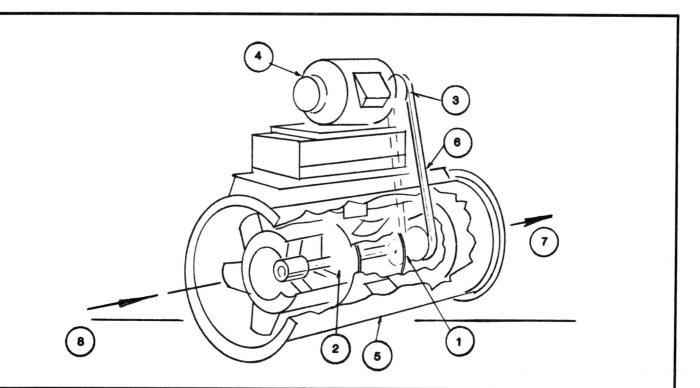

NO.	LOCATION OF POINTS DESCRIPTION	READINGS					
		HORIZONTAL		VERTICAL		AXIAL	
		VEL.	DISP.	VEL.	DISP.	VEL.	DISP.
1.	FAN BEARING, DRIVE END						
2.	FAN BEARING, OPPOSITE END						
3.	MOTOR BEARING, DRIVE END						
4.	MOTOR BEARING, OPPOSITE END						
5.	CASING (BOTTOM)						
6.	CASING (TOP)						
7.	DUCT AFTER FLEXIBLE CONNECTION (DISCHARGE)						
8.	DUCT AFTER FLEXIBLE CONNECTION (SUCTION)						

Normally Acceptable Readings: VEL. _____ ACC. _____

Unusual Conditions At Time Of Test: _____

Vibration Source (If Non Complying): _____

Remarks: _____

Form #82406

AABC LOGO

Date _____

Page _____ of _____

Project _____

System _____

DUCT LEAK TEST

Description of ductwork under test _____
_____
_____
_____
_____
_____

Duct Design Operating Static Pressure _____ " W.G.

Duct Design Test Static Pressure _____ " W.G.

Duct Capacity _____ CFM

Maximum Allowable Leakage: (Duct Capacity _____ CFM) (Leak Factor _____) = _____ CFM

Test Results

Test Apparatus

    Blower: _____

    Orifice: Tube Size _____ " Diameter   Orifice Size _____ " Diameter

    Calibrated: _____

Test Static Pressure: _____ " W.G.

Test Orifice Differential Pressure: _____ " W.G.   Leakage _____ CFM

Certified By: _____

Remarks: _____
_____
_____
_____

Form #82500

26.35
SECTION V—PROCEDURES

AABC LOGO

Date _____

Page _____ of _____

Job Name _____ Test Engineer _____

Boiler Make & Model _____ Serial No. _____

Burner Make & Model _____ Serial No. _____

COMBUSTION TEST

	GAS FIRING			OIL FIRING		
Test Number						
Firing Rate						
Overfire Draft (Inches WC)						
Last Pass Draft (Inches WC)						
Oil Pressure to Burner						
Gas Meter Timing Dial Size						
Gas Meter Time per Revolution						
Gas Pressure at Meter Outlet						
CFH Gas or Gal. Oil per Hr.						
Input MBH ____ BTU Gas / ____ BTU Oil						
Burner Manifold Gas Pressure (Inches WG)						
% Carbon Monoxide (CO)						
% Carbon Dioxide (CO_2)						
% Oxygen (O_2)						
% Excess Air						
Flue Gas Temp. at Outlet						
Ambient Temp.						
Net Stack Temp.						
% Stack Loss						
% Combustion Efficiency						
Output MBH						

Remarks: _____

Form #82600

CHAPTER 27

REPORT ANALYSIS PROCEDURES

27.1 OVERVIEW

Once the **Total System Balance** Report is submitted, the Engineer must analyze it to learn whether the system has achieved design intent.

An AABC Test and Balance Agency assures that the **Total System Balance** report is an accurate, professional record of the installation and operation of building systems. However, before a report can be analyzed it is necessary to understand what it is intended to represent. Often, a completely accurate report is challenged because the AABC Test and Balance Agency and the Engineer view it from two entirely different frames of reference.

This chapter describes the nature and intent of a **Total System Balance** report. It also discusses its uses, limitations, and how the Engineer can use it to check design intent.

27.2 WHAT A REPORT PROVIDES

An AABC **Total System Balance** report is:
A. Certified proof that the systems have been Tested and Balanced according to specifications within the limits of the installed systems.
B. An accurate representation of how the systems have been installed by the contractors.
C. A true representation of how the systems are operating at the completion of **Total System Balance**.
D. A record of all final quantities measured. These establish normal operating values of the systems.

27.3 USING A REPORT

When a **Total System Balance** report is submitted by an AABC Agency, it is a factual, accurate statement of the operating conditions of the systems at the completion of balancing. As a statement of actual operating conditions, the report can be used for three major functions:
A. Design evaluation
B. Verification of conditions
C. Data records

.1 Design Evaluation

The report depicts actual system operating conditions. It provides the Designer the opportunity to determine how well the project has met the original design intent. Design is the creative result of abstract ideas and calculations. The report is the concrete evidence of achievement of the design.

When used in this way, the report provides opportunity to improve design of future projects. The report can show design advantages and disadvantages, and can point out unanticipated job conditions that affected performance.

.2 Verification

The report is a verification document by:
A. Pointing out installation omissions and errors.
B. Verifying the work of contractors.
C. Pointing out conditions not in accordance with design documents.
D. Verifying the temperature control system.

.3 Data Records

The report is an invaluable, permanent record of the project. It provides:

A. System and equipment normal operating values that are of assistance to the Building Operators.
B. Data on equipment that allows Building Operators to order replacement parts.
C. The building static pressure.
D. A reference document for such future activities as: retrofit work, rebalancing, energy audits, troubleshooting, and maintenance.

27.4 INTERPRETING REPORTS

.1 Field and Laboratory Conditions

It is important to understand that there is a great difference between field data and manufacturer's ratings. The report is the result of field-gathered data, and the data will seldom be the same as manufacturer's ratings which were calculated from laboratory procedures. Some reasons are as follows:

A. The laboratory has a controlled environment which is designed for testing.
B. The configurations of connections to equipment in the laboratory are designed for accurate measurement. In the field they are not.
C. Laboratory and field test equipment are designed for different intents and priorities.
D. The application of the instrumentation is different in the laboratory than it is in the field.
E. There is a difference between laboratory and field personnel; and between their objectives.
F. Parameters (such as electrical values) are carefully controlled in the laboratory. They may vary in the field.

There are also differences in manufacturer's ratings and measured field data because:

A. Manufacturer's data is not always based upon proven laboratory procedures. Often it is calculated from engineering factors.
B. Fan and pump ratings, as determined in the laboratory, are modified by the "SYSTEM EFFECT." This is the unmeasurable effect that the method of connection to the system has upon a fan or pump. For example, a fan with poorly designed inlet and outlet connections will have greatly reduced performance because of System Effect, which cannot be measured. However, it can be calculated. One reference for more information is "FANS AND SYSTEMS," AMCA Publication 201.

.2 Equipment Data

Equipment performance data recorded in the **Total System Balance** report is often questioned because calculations based upon manufacturer's ratings indicate substantially different values. There are many reasons for the variation between actual performance and calculated performance—all based upon the difference between field conditions and laboratory conditions. The following are some reasons for major pieces of equipment.

A. Fans—Generally, the manufacturer's fan tables reflect fan performance under ideal conditions—not under the unfavorable conditions sometimes encountered on the job. Also, the system characteristics often make a fan operate to the left of peak performance on the fan curve. The result is that a static pressure reading may not result in the CFM shown on the fan curve at the point to the right of peak performance. Instead, it results in the lower CFM shown by the corresponding point that is to the left of peak performance.
B. Pumps—The static pressure rise, as measured in the field cannot be used to determine pump capacity for the follow-

ing reasons:
1. The pump is not a flow measuring device.
2. This method would ignore the Velocity Head Component.
3. Pressure taps are not located in the same place as they were under-rated test conditions.
4. Piping arrangements are not the same as rated test conditions.
5. The rating of the installed pump generally has not been laboratory confirmed.

C. Coils—Manufacturer's coil ratings may have been calculated rather than laboratory tested. This means that measured field data may vary greatly from manufacturer's ratings. In addition, if the ratings are from laboratory measurements, field operating conditions for coils are different from those in the laboratory. Also, the physical condition of the coil may differ from the tested conditions for the following probable reasons.
1. Airflow across the coil is not uniform.
2. Coil fins are not completely straight.
3. Fins are not completely bonded to the tube.
4. Foreign material is on the fins of the coil.
5. Fouling occurring in the tubes.
6. Air temperatures entering and leaving the coil are stratified and uneven.
7. Leakage occurring between the duct and the coil.

D. Chillers—**Total System Balance** procedures are only intended to establish normal operating values. Chiller performance testing is not a function of **Total System Balance**.

E. Hydronic Flow Meters—Field conditions are such that the accuracy of the meters may be quite different from that determined in the laboratory. Their location in the system can also affect their accuracy. (For example, too close to a coil or a pump.) The gages used with the meters will be field type—which are not as accurate as laboratory gages. However, these field-type gages provide more than sufficient accuracy for the purpose of proportioning flow quantities.

F. Electrical—It is important to recognize that the measured field data of amperes and voltage for a motor **cannot be used** to determine accurate brake horsepower. The reason is that necessary data for the calculations are seldom known.
1. The power factor is generally not measured.
2. Motor efficiency, and drive efficiency are generally not available. (See Chapter 10 of the AABC National Standards, 1982, for methods of calculating brake horsepower.)

G. Pressure Differential vs. Flow Rate

Often the field measured differential pressure across components does not agree with manufacturer's published data for the flow quantity that was measured. This does not infer that the measured data is wrong or that the published information is inaccurate. The difference is usually because the entrance and leaving conditions in the field are not the same as the test conditions under which the component was rated.

CHAPTER 28

REPORT VERIFICATION PROCEDURES

28.1 OVERVIEW

This chapter describes the Standards by which the Owner's Representative verifies that the **Total System Balance** has been performed according to the AABC National Standards, 1982, and to the project specifications.

Two types of verification are described:
- Progress Field Inspections
- Report Verification Procedures (Office and Field)

Verification procedures are essential to protect the Owner from improper **Total System Balance**. These Verification Procedures will compel any firm not performing to the AABC Standards to repeat completed Testing and Balancing work until it meets acceptable quality standards.

28.2 PURPOSE

The purpose of Progress Field Inspections and Report Verification Procedures is to verify:
- The integrity of the Report.
- That the work has been performed in accordance with the AABC National Standards, 1982, and in accordance with the project specifications.

28.3 PROGRESS FIELD INSPECTIONS

Progress Field Inspections should be performed while **Total System Balance** work is in progress. The Owner's Representative should witness actual **Total System Balance** work. In particular, any tests which are of special concern to the Owner should be witnessed.

Familiarity with the AABC National Standards, 1982; with the project specifications; and with **Total System Balance** procedures in general are essential for the Owner's Representative to recognize that work is being performed to required quality standards.

28.4 REPORT VERIFICATION PROCEDURES

Report Verification Procedures must be performed within 30 days after the **Total System Balance** Report has been received.

.1 Office Verification Procedures

The Owner's Representative should perform the following Verification Procedures prior to Field Verification:

A. Review the **Total System Balance** Report.

B. Check that the Report reflects all Addenda and Change Orders. The Report must be based on the latest Contract Documents.

It should be recognized that measured data may vary from design data (within reasonable limits).

.2 Field Verification Procedures

A. The Owner's Representative should select 10% of the report data for verification. Selection should be at random. "Report data" is defined as one tabulated item on a report form, such as the air velocity at a specific outlet; air or water flow quantity; differential pressure reading; or electrical or sound measurement.

B. The Test and Balance Agency shall be given sufficient advance notice of the date of Field Verification. However, the Test and Balance Agency is not to be in-

Chapter 28—Report Verification Procedures

formed in advance of the data to be verified.

C. The Test and Balance Agency must have the right to set any system mode to Report data conditions.
D. The Test and Balance Agency must use the same instrument (by Model and Serial Number) that was used when the original data was read and recorded.

.3 Field Verification Standards

A. For Field Verification readings to be considered valid they must be witnessed by the Owner's Representative.
B. Failure of an Item
   1. For all readings other than sound, a deviation of more than 10% between the Verification Reading and the Reported Data shall be considered as failing the Verification Procedure for that item.
   2. A deviation of 3 decibels in sound pressure levels between the Sound Verification Reading and the Sound Report Data shall be considered as failing the Verification Procedure for that item. Variations in background noise must be considered.
C. A failure of more than 10% of the selected items shall result in the failure of the entire Field Verification Procedure.

.4 Failure of Field Verification Procedure

For a project that fails to meet Field Verification Standards as stated in 28.4.3, the **Total System Balance** Agency must perform the following work at no additional cost:

A. Any system failing the Verification Procedure must be re-balanced.
B. New **Total System Balance** Reports must be provided.
C. New Field Verification Procedures must be performed.

The specifications in this chapter are interrelated with those in other chapters of this section and must be properly combined to form a complete set of specifications. See Chapter 16 for a checklist of elements to form a complete set of specifications.

APPENDIX

**AABC
NATIONAL STANDARDS

1982**

AABC TECHNICAL REPORT

COOLING TOWER FIELD TESTING AND CONVERSION DATA

OVERVIEW:

To provide a complete total System Balance it is necessary to do a cooling tower test to determine its operating capacity as installed. Section 23.6 of the AABC National Standards, 1982, details a complete field test procedure and the necessary calculations for the consulting engineer or the test and balance engineer to determine the actual performance of an installed tower that is tested.

The field test procedure described in this AABC Technical Report provides an accurate test of a cooling tower performance at ambient conditions and heat load conditions in the building that vary from the design conditions. Since a cooling tower test can seldom if ever, be made at both the design heat load and the design entering wet bulb (WB) condition, it is desirable to have an accurate test procedure that can be related to the design requirements. This test procedure provides such an accurate and practical approach.

GENERAL:

The objective of this test is to determine a performance factor (P.F.) for the tower that is constant at all heat load conditions and all entering (WB) wet bulb conditions that may occur. The performance factor (P.F.) will be used to determine the performance of the tower at the design conditions.

Following is an outline of the analysis that establishes this test procedure, according to applicable cooling tower factors and how they operate:

1) Water flow (GPM) and air flow across the tower are constant.

2) At constant heat load, the approach will decrease as the entering WB increases. Approach is defined as the difference between the temperature of the water entering (on) the tower and the wet bulb temperature of the air entering the tower.

3) Evaporation takes place in a tower but has no affect on the cooling of the water. It is simply a conversion of air sensible heat to latent heat. Test shows the same tower capacity with air entering at 95° dry bulb (DB) and 78° WB as they do at 79° WB for the same tower under the same load.

4) For standard cooling tower, the heat rejected per ton of refrigeration is 15,000 BTU/Hr.

5) At any point of test, the water, air and fill arrangement combine to allow the air to remove a given percentage of the theoretical heat available to it. This percentage, called the Performance Factor (P.F.), will remain constant at any entering WB temperature and heat load as long as the GPM, CFM and fill do not change.

TEST PROCEDURE:

During the test procedure, the following temperatures are recorded:

1) Water temperature entering the tower.

2) Water temperature leaving the tower.

3) WB temperature of the air entering the tower.

4) WB temperature of the air leaving the tower.

Appendix

It is also necessary to know the volume of water that is being circulated through the tower. Since this flow rate is constant, it need not be measured during the test period.

For details of the measurement technique and the degree of accuracy that is required, refer to Section 23.6 in the AABC National Standards, 1982.

ANALYSIS OF THE TEST DATA:

The enthalpy (total heat) of the air at the leaving WB temperature minus the enthalpy of the air at the entering WB temperature determines the quantity of heat in BTU/Lb. that the air picks up while passing through the tower. If this heat gain (enthalpy difference), is divided by the theoretical enthalpy difference that would result if the air left the tower at a wet bulb temperature equal to the temperature of the water entering the tower, the result is the Performance Factor. This theoretical enthalpy difference occurs at zero approach and represents a tower performance of 100%.

$$\frac{\text{Measured enthalpy difference}}{\text{Theoretical enthalpy difference}} =$$

Performance Factor (P.F.)

If the water flow (GPM) and the air flow through the tower remain constant, then the P.F. will remain the same as the heat load and WB temperature change.

With this as a basis, a cooling tower can be tested at a reasonable operating condition and from the data of this test calculate the performance of the tower at the specified full load conditions. It is possible to prepare a performance curve for the test tower at a given P.F. It has been found that the most useful curve is to plot approach vs. heat rejection at specified entering WB temperature, test water flow (GPM) and test P.F. From this curve, the engineer can determine if the condenser water system will satisfy the requirements of the operation of the tower at a whole range of operating conditions.

COOLING TOWER TERMS

EWT = Water temperature on tower °F.
LWT = Water temperature off tower °F.
EWB = Air entering tower wet bulb °F.
LWB = Air leaving tower wet bulb °F.
GPM = Gallons of water per minute over tower at test.
d = Air density in Lbs. per cubic ft. leaving tower.
R = Range across tower = EWT − LWT.
A = Approach to wet bulb = LWT − EWB.
h = Enthalpy of air at a wet bulb temperature of EWT.
h_2 = Enthalpy of air at LWB.
h_3 = Enthalpy of air at EWB.
PF = Tower performance capacity factor. This is a percentage of the heat picked up by the air in passing through the tower to the amount it would have picked up if it had left the tower saturated at the EWT.

The Equation is: $PF = \dfrac{h_2 - h_3}{h - h_3}$

THR = Total heat rejected by tower in BTU per minute
= GPM × 8.33 × (EWT − LWT)
Lam = Lbs. of air per minute through tower
$= \dfrac{THR}{h_2 - h_3}$
CFM = Tower cfm $= \dfrac{Lam}{d \text{ (of leaving air)}}$

The simple way to explain our method is to take an example of a tower specification and the test results.

	SPECIFIED	TEST
GPM	500	490
Water on Tower	95	91
Water off Tower	85	83
Entering Wet Bulb	79	75
Leaving Air Wet Bulb	—	86

FROM TEST DATA

1. Find Tower test PF from test data as follows:

   $h_2 - h_3 =$

   Enthalpy of LWB =

   $\quad\quad 86°F = \quad 50.66$

   Enthalpy of EWB =

   $\quad\quad 75°F = \quad \underline{38.61}$

   $\quad\quad\quad\quad\quad\quad 12.05$ BTU/Lb.

   $h - h_3 =$

   Enthalpy of air @ EWT =

   $\quad\quad 91°F = \quad 57.33$

   Enthalpy of air @ EWB =

   $\quad\quad 75°F = \quad \underline{38.61}$

   $\quad\quad\quad\quad\quad\quad 18.72$ BTU/Lb.

   Tower PF = 12.05/18.72 = .640

2. Find total heat rejected at test (THR):

   R = 91 - 83 = 8°F

   Test THR = 490 × 8.33 × 8 = 32,650 BTU/Min.

3. Find Lbs. of air/Min. through Tower (Lam):

   Lam = THR/$h_2 - h_3$ =

   32,650/12.05 = 2,705 Lbs./Min.

TO CONVERT TO SPECIFIED DATA

1. Specified THR = 500 × 8.33 × (95 - 85) = 41,600 BTU/Min.

2. Required Range (R) with this pump =

   $\dfrac{41,600}{490 \times 8.33} = 10.2°F$

3. Required $h_2 - h_3$ at design =

   $\dfrac{41,600}{2,705} = 15.35$ BTU/lb.

   (Lam from test)

4. Find EWT as follows:

   $\dfrac{h_2 - h_3}{PF}$ + Enthalpy of design EWB =

   $\dfrac{15.35}{.64}$ + 42.62 =

   24.00 + 42.62 = 66.62 BTU/lb. =

   wet bulb temperature of 97.05°F = EWT.

5. Find leaving water temperature at specified conditions as follows:

   EWT - R = 97.05 - 10.20 = 86.85°F.

6. Then at specified load and wet bulb temperatures the Tower will cool 490 GPM from 97.05° to 86.85° at 79.0°F entering wet bulb.

7. Specified approach = 85 - 79 = 6°F

   Actual approach = 86.85 - 79 = 7.85°F

8. Prepare a graph for actual approach (°F) versus the total heat rejected in BTU per minute. At the specified total heat rejection of 41,600 BTU/Min., this data becomes one point on the curve. In the same manner, the actual approach is calculated for the total heat rejection at the rate of 32,650 BTU/Min., and at three other rates of heat rejection: 35,000, 30,000, and 25,000 BTU/Min.

The resulting points are plotted and the curve that is drawn represents a test curve at 79°F/W.B. for the tower under test.

The specified approach for this tower is 6.0°F and the specified rate of total heat rejection is 41,600 BTU/Min. The test curve indicates that at the specified approach, the rate of heat rejection will be 26,800 BTU/Min. This tower does not meet the specified capacity requirements.

A.4 Appendix

TOWER TEST CURVE AT 79°F APPROACH VS. HEAT REJECTION

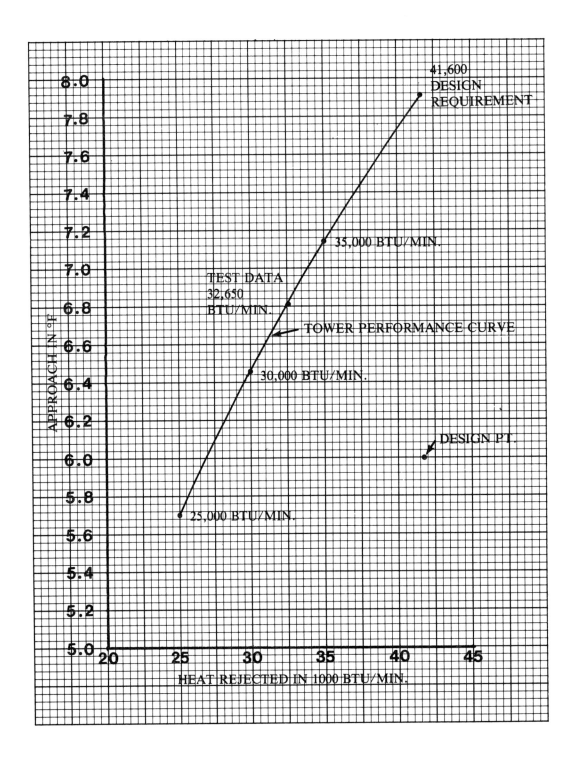

VELOCITY PRESSURE vs. VELOCITY

Velocity pressure in inches of water column
Velocity in feet per minute
Standard air density of .07495 lbs./ft.^{3}

VP	V	VP	V	VP	V	VP	V	VP	V	VP	V
.001	127	.026	645	.051	904	.076	1104	.101	1273	.126	1422
.002	179	.027	658	.052	913	.077	1111	.102	1279	.127	1427
.003	219	.028	670	.053	922	.078	1119	.103	1285	.128	1433
.004	253	.029	682	.054	931	.079	1125	.104	1292	.129	1439
.005	283	.030	694	.055	939	.080	1133	.105	1298	.130	1444
.006	310	.031	705	.056	948	.081	1140	.106	1304	.131	1449
.007	335	.032	716	.057	956	.082	1147	.107	1310	.132	1455
.008	358	.033	727	.058	964	.083	1154	.108	1316	.133	1461
.009	380	.034	738	.059	973	.084	1161	.109	1322	.134	1466
.010	400	.035	749	.060	981	.085	1167	.110	1328	.135	1471
.011	420	.036	759	.061	989	.086	1175	.111	1334	.136	1477
.012	439	.037	770	.062	996	.087	1181	.112	1340	.137	1482
.013	457	.038	780	.063	1004	.088	1188	.113	1346	.138	1488
.014	474	.039	791	.064	1012	.089	1193	.114	1352	.139	1493
.015	491	.040	801	.065	1020	.090	1201	.115	1358	.140	1498
.016	507	.041	811	.066	1029	.091	1208	.116	1364	.141	1504
.017	522	.042	821	.067	1037	.092	1215	.117	1370	.142	1509
.018	537	.043	831	.068	1045	.093	1222	.118	1376	.143	1515
.019	552	.044	840	.069	1052	.094	1228	.119	1382	.144	1520
.020	566	.045	849	.070	1060	.095	1234	.120	1387	.145	1525
.021	580	.046	859	.071	1067	.096	1241	.121	1393	.146	1530
.022	594	.047	868	.072	1075	.097	1247	.122	1399	.147	1536
.023	607	.048	877	.073	1082	.098	1254	.123	1404	.148	1541
.024	620	.049	887	.074	1089	.099	1260	.124	1410	.149	1546
.025	633	.050	895	.075	1097	.100	1266	.125	1416	.150	1551

PERMISSION KAHOE AIR BALANCE COMPANY

VELOCITY PRESSURE vs. VELOCITY Contd.

VP	V	VP	V	VP	V	VP	V	VP	V	VP	V
.151	1556	.176	1680	.21	1835	.46	2716	.71	3375	.96	3924
.152	1561	.177	1685	.22	1879	.47	2746	.72	3398	.97	3945
.153	1567	.178	1690	.23	1921	.48	2775	.73	3422	.98	3965
.154	1572	.179	1695	.24	1962	.49	2804	.74	3445	.99	3985
.155	1577	.180	1699	.25	2003	.50	2832	.75	3468	1.00	4005
.156	1582	.181	1704	.26	2042	.51	2860	.76	3491	1.01	4025
.157	1587	.182	1709	.27	2081	.52	2888	.77	3514	1.02	4045
.158	1592	.183	1713	.28	2119	.53	2916	.78	3537	1.03	4064
.159	1597	.184	1718	.29	2157	.54	2943	.79	3560	1.04	4084
.160	1602	.185	1723	.30	2193	.55	2970	.80	3582	1.05	4103
.161	1607	.186	1727	.31	2230	.56	2997	.81	3604	1.06	4123
.162	1612	.187	1732	.32	2260	.57	3024	.82	3625	1.07	4142
.163	1617	.188	1737	.33	2301	.58	3050	.83	3657	1.08	4162
.164	1622	.189	1741	.34	2335	.59	3076	.84	3669	1.09	4181
.165	1627	.190	1746	.35	2369	.60	3102	.85	3690	1.10	4200
.166	1632	.191	1750	.36	2403	.61	3127	.86	3709	1.11	4219
.167	1637	.192	1755	.37	2436	.62	3153	.87	3729	1.12	4238
.168	1642	.193	1759	.38	2469	.63	3179	.88	3758	1.13	4257
.169	1646	.194	1764	.39	2501	.64	3204	.89	3779	1.14	4276
.170	1651	.195	1768	.40	2533	.65	3229	.90	3800	1.15	4295
.171	1656	.196	1773	.41	2563	.66	3254	.91	3821	1.16	4314
.172	1661	.197	1777	.42	2595	.67	3279	.92	3842	1.17	4332
.173	1666	.198	1782	.43	2626	.68	3303	.93	3863	1.18	4350
.174	1670	.199	1787	.44	2656	.69	3327	.94	3884	1.19	4368
.175	1675	.200	1791	.45	2687	.70	3351	.95	3904	1.20	4386

PERMISSION KAHOE AIR BALANCE COMPANY

AIR VELOCITY AGAINST VELOCITY PRESSURE FOR STANDARD AIR

$$\text{VELOCITY} = 4005 \sqrt{\text{VELOCITY PRESSURE}}$$

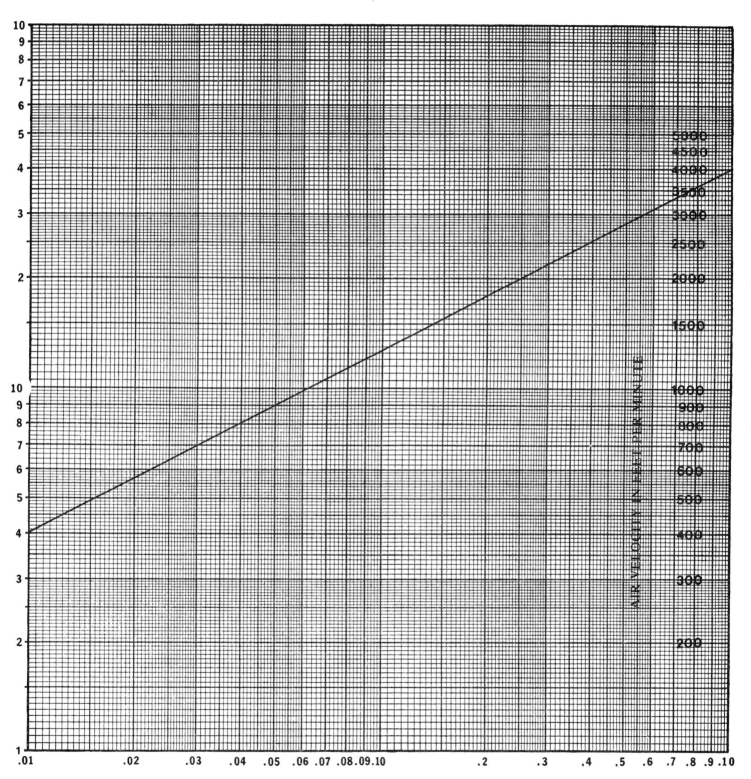

VELOCITY PRESSURE IN INCHES OF WATER

PERMISSION KAHOE AIR BALANCE COMPANY

A.8 Appendix

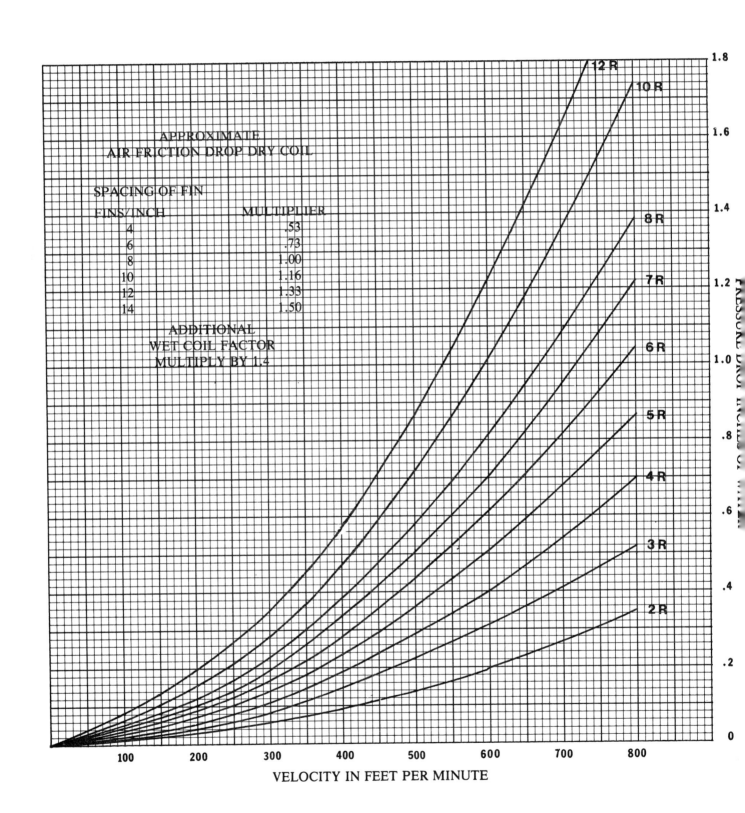

PERMISSION KAHOE AIR BALANCE COMPANY

ΔT_A — AIR TEMPERATURE DIFFERENCE = $\dfrac{BTU}{1.08 \times CFM}$

ΔT_W — WATER TEMPERATURE DIFFERENCE = $\dfrac{MBH \times 2}{GPM}$

$GPM = \dfrac{MBH \times 2}{\Delta T_W}$ $\qquad MBH = \dfrac{\Delta T_W \times GPM}{2}$

$BTU = 1.08 \times CFM \times \Delta T_A$

(MBH = 1000 BTU/HR)

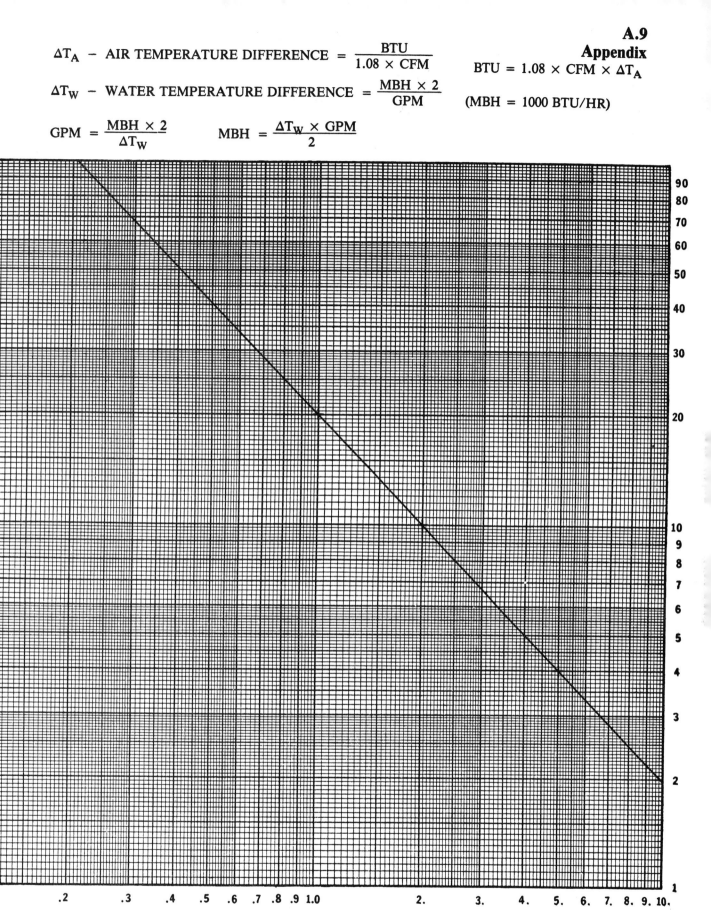

WATER TEMPERATURE DROP VS GPM

PERMISSION KAHOE AIR BALANCE COMPANY

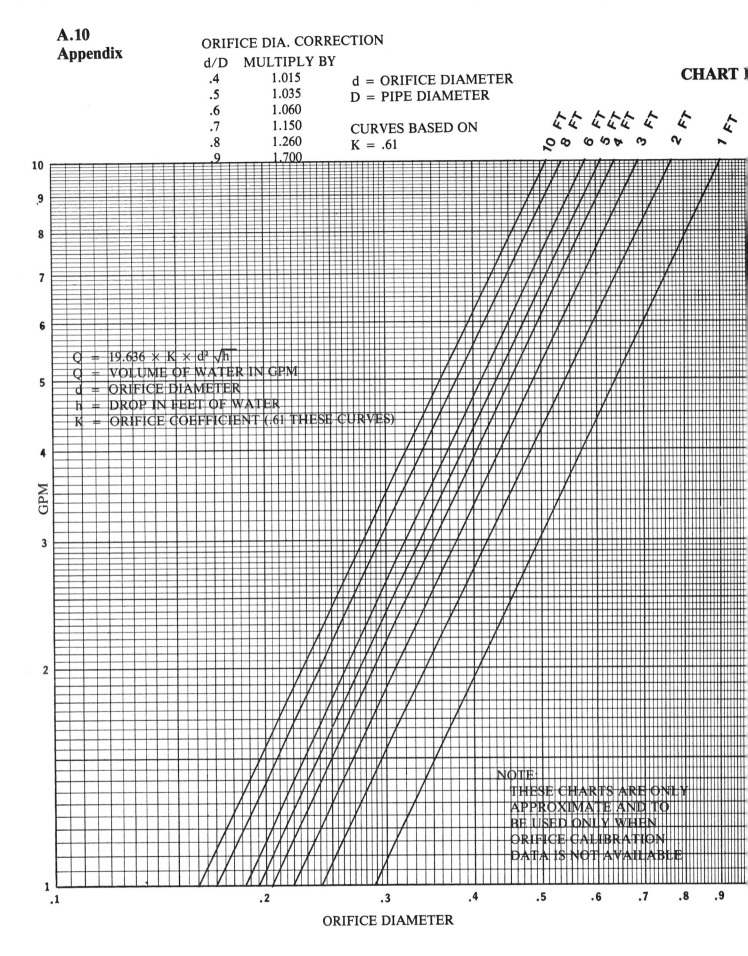

ORIFICE PRESSURE VS GPM (WATER)

A.11
Appendix

CHART II

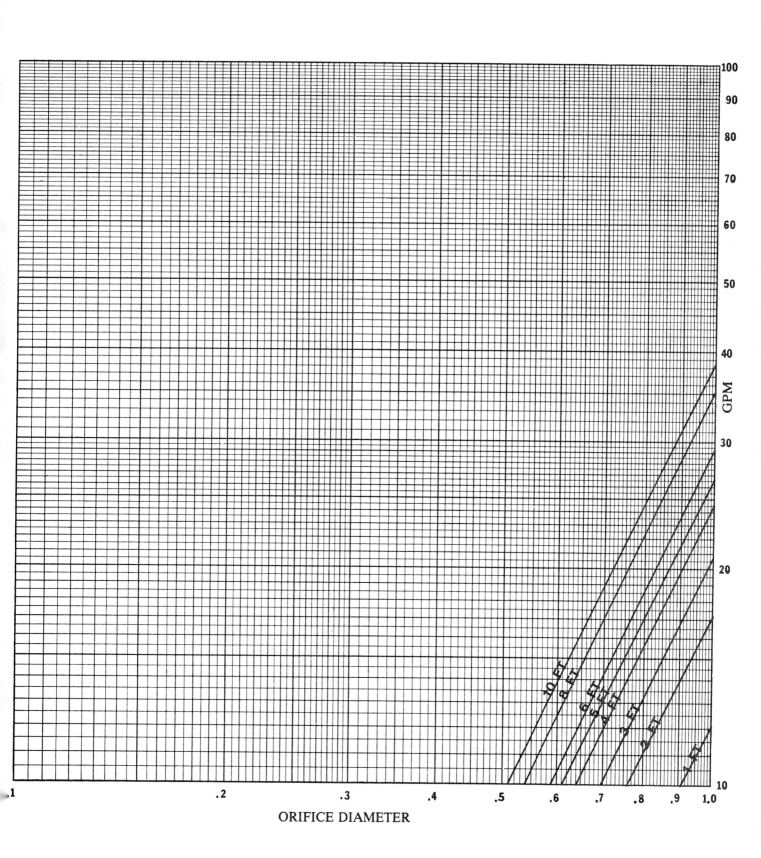

ORIFICE PRESSURE VS GPM (WATER)

PERMISSION KAHOE AIR BALANCE COMPANY

A.12
Appendix

CHART III

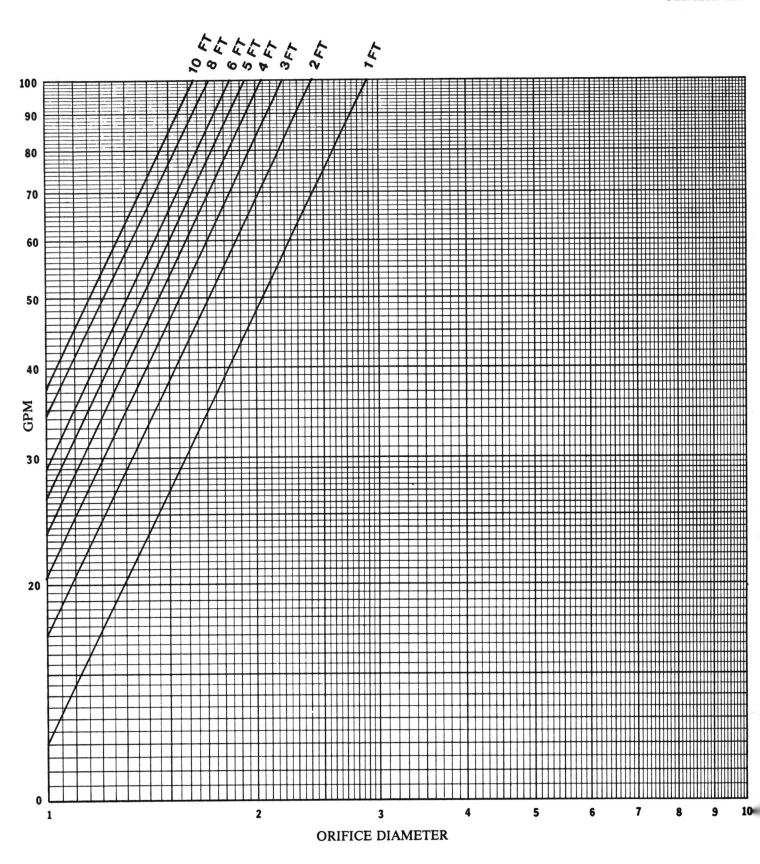

ORIFICE PRESSURE VS GPM (WATER)

PERMISSION KAHOE AIR BALANCE COMPANY

A.13
Appendix
CHART IV

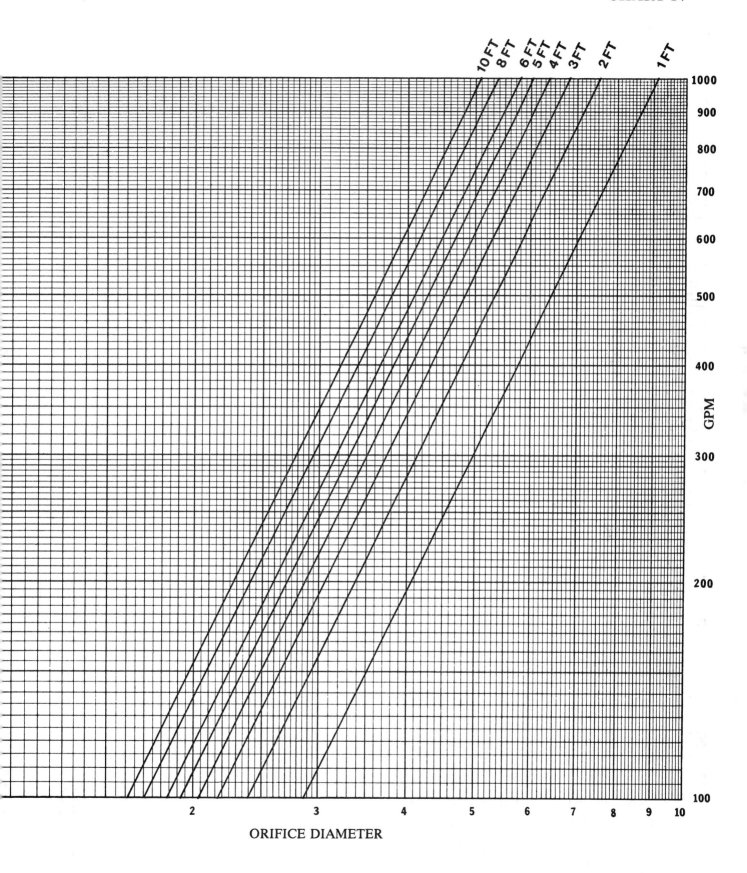

ORIFICE PRESSURE VS GPM (WATER)

PERMISSION KAHOE AIR BALANCE COMPANY

A.14
Appendix

HEAT TRANSFER—WATER #1

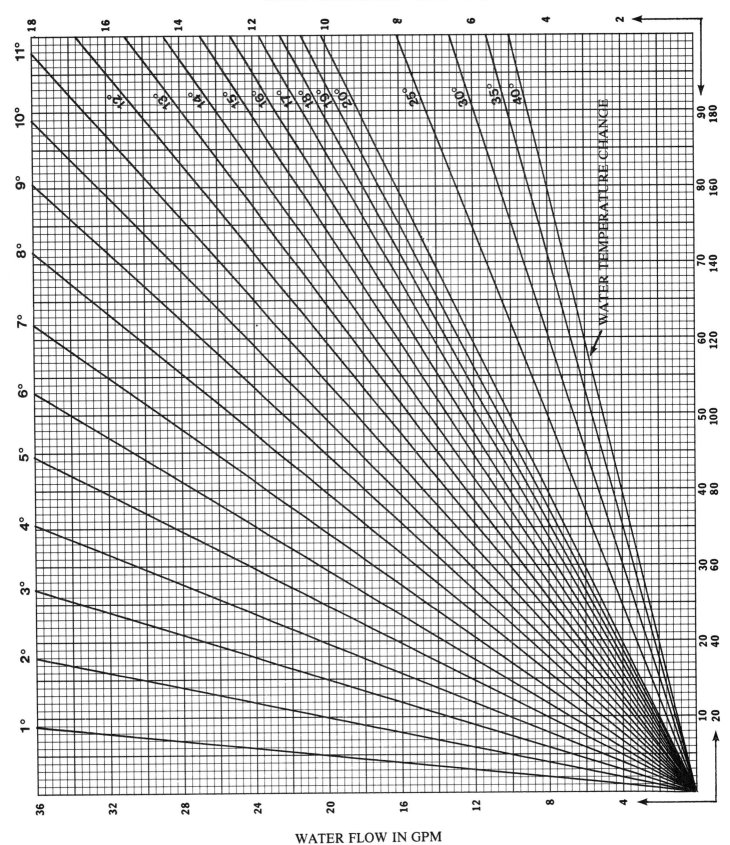

PERMISSION KAHOE AIR BALANCE COMPANY

HEAT TRANSFER—WATER #2

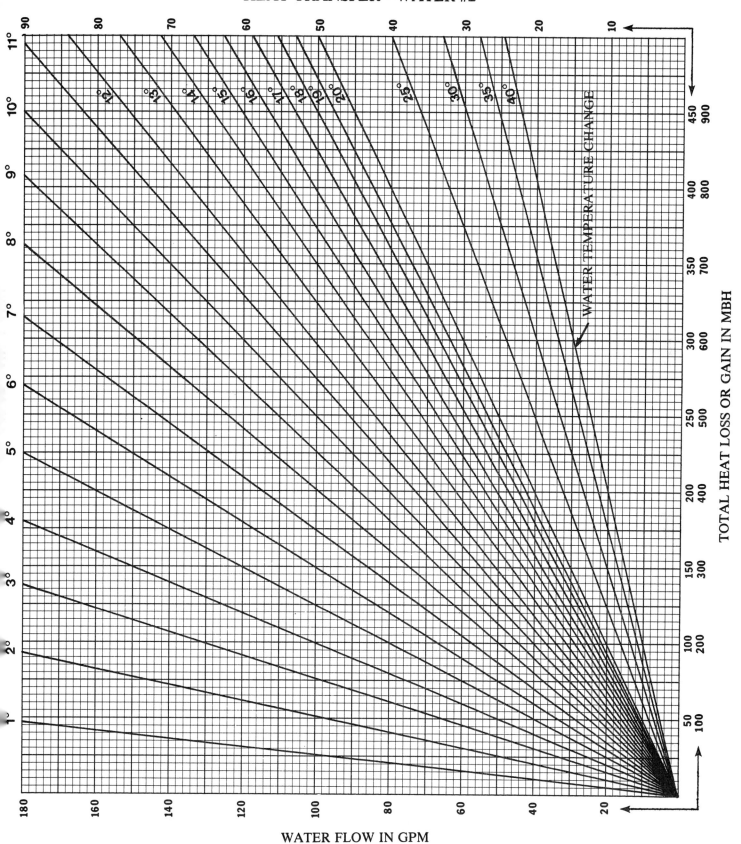

PERMISSION KAHOE AIR BALANCE COMPANY

A.15
Appendix

A.16 Appendix

HEAT TRANSFER—WATER #3

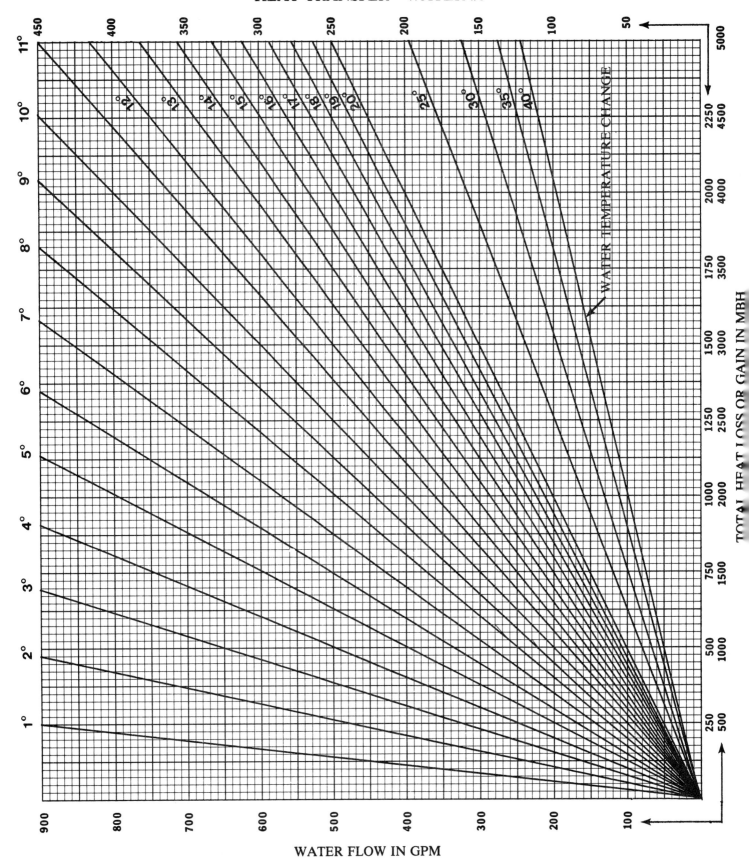

PERMISSION KAHOE AIR BALANCE COMPANY

ENTHALPY TABLE
STANDARD AIR
BTUH/CFM at 29.92 IN. HG.

WB Temp.	0	.1	.2	.3	.4	.5	.6	.7	.8	.9
34	56.63	56.82	57.01	57.20	57.39	57.58	57.77	57.96	58.16	58.35
35	58.54	58.73	58.92	59.12	59.31	59.50	59.70	59.89	60.08	60.28
36	60.47	60.67	60.86	61.06	61.26	61.45	61.65	61.84	62.04	62.24
37	62.43	62.63	62.83	63.03	63.23	63.43	63.63	63.83	64.04	64.24
38	64.44	64.64	64.84	65.05	65.25	65.45	65.66	65.86	66.06	66.27
39	66.47	66.68	66.88	67.09	67.30	67.50	67.71	67.92	68.12	68.33
40	68.54	68.75	68.96	69.17	69.38	69.59	69.80	70.01	70.22	70.43
41	70.64	70.85	71.06	71.28	71.49	71.71	71.92	72.13	72.35	72.56
42	72.77	72.99	73.21	73.43	73.65	73.87	74.08	74.30	74.52	74.74
43	74.96	75.18	75.40	75.62	75.84	76.06	76.28	76.51	76.73	76.95
45	79.42	79.65	79.88	80.11	80.34	80.57	80.80	81.03	81.26	81.49
46	81.72	81.96	82.19	82.43	82.66	82.89	83.13	83.36	83.59	83.83
47	84.06	84.30	84.54	84.78	85.02	85.25	85.49	85.73	85.97	86.21
48	86.45	86.69	86.94	87.18	87.42	87.66	87.91	88.15	88.39	88.64
49	88.88	89.13	89.37	89.62	89.87	90.12	90.36	90.61	90.86	91.11
50	91.35	91.61	91.86	92.11	92.36	92.62	92.87	93.12	93.37	93.63

PERMISSION KAHOE AIR BALANCE COMPANY

ENTHALPY TABLE
STANDARD AIR
BTUH/CFM at 29.92 IN. HG.

WB Temp.	0	.1	.2	.3	.4	.5	.6	.7	.8	.9
51	93.88	94.14	94.40	94.65	94.91	95.17	95.43	95.69	95.95	96.20
52	96.46	96.72	96.99	97.25	97.51	97.78	98.04	98.30	98.56	98.83
53	99.09	99.36	99.63	99.89	100.16	100.43	100.70	100.96	101.23	101.50
54	101.77	102.04	102.31	102.58	102.86	103.13	103.40	103.67	103.95	104.22
55	104.49	104.77	105.05	105.33	105.61	105.88	106.16	106.44	106.72	107.00
56	107.28	107.57	107.86	108.14	108.43	108.72	109.01	109.30	109.58	109.87
57	110.16	110.45	110.74	111.02	111.31	111.60	111.89	112.18	112.46	112.75
58	113.04	113.34	113.63	113.93	114.23	114.52	114.82	115.12	115.42	115.71
59	116.01	116.32	116.62	116.93	117.23	117.54	117.85	118.15	118.46	118.76
60	119.07	119.38	119.69	120.00	120.31	120.62	120.93	121.24	121.55	121.86
61	122.18	122.49	122.81	123.12	123.44	123.75	124.07	124.38	124.70	125.01
62	125.33	125.65	125.97	126.30	126.62	126.95	127.27	127.59	127.92	128.24
63	128.57	128.90	129.23	129.56	129.90	130.23	130.56	130.90	131.23	131.56
64	131.90	132.23	132.57	132.91	133.25	133.58	133.92	134.26	134.60	134.93
65	135.27	135.62	135.96	136.31	136.66	137.00	137.35	137.70	138.04	138.39

PERMISSION KAHOE AIR BALANCE COMPANY

A.19 Appendix

TABLE VIII — BTUH/CFM DIFFERENTIAL TABLE — TOTAL HEAT TRANSFER

INITIAL WET BULB TEMPERATURE

Final Wet Bulb Temp.	60	61	62	63	64	65	66	67	68	69	70	71	72	73	74	75	76	77	78	79	80	Final Wet Bulb Temp.
40	50.5	53.6	56.8	60.0	63.4	66.7	70.2	73.8	77.4	81.1	84.9	88.7	92.7	96.8	100.9	105.2	109.5	114.0	118.6	123.3	128.1	40
41	48.4	51.5	54.7	57.9	61.2	64.6	68.1	71.6	75.2	79.0	82.8	86.6	90.6	94.7	98.8	103.1	107.4	111.9	116.5	121.1	126.0	41
42	46.3	49.4	52.6	55.8	59.1	62.5	66.0	69.5	73.1	76.9	80.6	84.5	88.5	92.6	96.7	101.0	105.6	109.8	114.3	119.0	123.8	42
43	44.1	47.2	50.4	53.6	56.9	60.3	63.8	67.3	70.9	74.7	78.4	82.3	86.3	90.4	94.5	98.8	103.1	107.6	112.1	116.8	121.6	43
44	41.9	45.0	48.2	51.4	54.7	58.1	61.6	65.1	68.7	72.5	76.2	80.1	84.1	88.2	92.3	96.6	100.9	105.4	109.9	114.6	119.4	44
45	39.6	42.8	45.9	49.1	52.5	55.8	59.3	62.9	66.5	70.2	74.0	77.9	81.8	85.9	90.0	94.3	98.6	103.1	107.7	112.4	117.2	45
46	37.4	40.5	43.6	46.8	50.2	53.6	57.0	60.6	64.2	67.9	71.7	75.6	79.5	83.6	87.8	92.0	96.3	100.8	105.4	110.1	114.9	46
47	35.0	38.1	41.3	44.5	47.8	51.2	54.7	58.2	61.8	65.6	69.3	73.2	77.2	81.3	85.4	89.7	94.0	98.5	103.1	107.7	112.5	47
48	32.6	35.7	38.9	42.1	45.5	48.8	52.3	55.8	59.4	63.2	67.0	70.8	74.8	78.9	83.0	87.3	91.6	96.1	100.7	105.3	110.2	48
49	30.2	33.3	36.5	39.7	43.0	46.4	49.9	53.4	57.0	60.8	64.5	68.4	72.4	76.5	80.6	84.9	89.2	93.7	98.2	102.9	107.7	49
50	27.7	30.8	34.0	37.2	40.5	43.9	47.4	50.9	54.5	58.3	62.1	65.9	69.9	74.0	78.1	82.4	86.7	91.2	95.8	100.4	105.3	50
51	25.2	28.3	31.5	34.7	38.0	41.4	44.9	48.4	52.0	55.8	59.5	63.4	67.4	71.5	75.6	79.9	84.2	88.7	93.2	97.9	102.7	51
52	22.6	25.7	28.8	32.1	35.4	38.8	42.3	45.8	49.4	53.1	56.9	60.8	64.8	68.9	73.0	77.3	81.6	86.1	90.6	95.3	100.1	52
53	20.0	23.1	26.2	29.5	32.8	36.2	39.6	43.2	46.8	50.5	54.3	58.2	62.1	66.2	70.4	74.7	79.0	83.5	88.0	92.7	97.5	53
54	17.3	20.4	23.5	26.8	30.1	33.5	36.9	40.5	44.1	47.8	51.6	55.5	59.4	63.5	67.7	72.0	76.3	80.8	85.3	90.0	94.8	54
55	14.6	17.7	20.8	24.2	27.4	30.8	34.2	37.8	41.4	45.1	48.9	52.8	56.7	60.8	65.0	69.3	73.6	78.1	82.6	87.3	92.1	55
56	11.8	14.9	18.0	21.3	24.6	28.0	31.5	35.0	38.6	42.3	46.1	50.0	54.0	58.1	62.2	66.5	70.8	75.3	79.8	84.5	89.3	56
57	8.9	12.0	15.2	18.4	21.7	25.1	28.6	32.1	35.7	39.5	43.2	47.1	51.1	55.2	59.3	63.6	67.9	72.4	77.0	81.6	86.4	57
58	6.0	9.1	12.3	15.5	18.9	22.2	25.7	29.3	32.9	36.6	40.4	44.2	48.2	52.3	56.4	60.7	65.0	69.5	74.1	78.8	83.6	58
59	3.1	6.2	9.3	12.6	15.9	19.3	22.7	26.3	29.9	33.6	37.4	41.3	45.2	49.7	53.5	57.7	62.1	66.6	71.1	75.8	80.6	59
60		3.1	6.3	9.5	12.8	16.2	19.7	23.2	26.8	30.6	34.3	38.2	42.2	46.3	50.4	54.7	59.0	63.5	68.0	72.7	77.5	60
61			3.2	6.4	9.7	13.1	16.6	20.1	23.7	27.5	31.2	35.1	39.1	43.2	47.3	51.6	55.9	60.4	64.9	69.6	74.4	61
62				3.2	6.6	9.9	13.4	17.0	20.6	24.3	28.1	32.0	35.9	40.0	44.1	48.4	52.7	57.2	61.8	66.5	71.3	62
63					3.3	6.7	10.2	13.7	17.3	21.1	24.8	28.7	32.7	36.8	40.9	45.2	49.5	54.0	58.5	63.2	68.0	63
64						3.4	6.8	10.4	14.0	17.7	21.5	25.4	29.3	33.4	37.6	41.9	46.2	50.7	55.2	59.9	64.7	64
65							3.5	7.0	10.6	14.4	18.1	22.0	26.0	30.1	34.2	38.5	42.8	47.3	51.8	56.5	61.3	65
66								3.6	7.2	10.9	14.7	18.5	22.5	26.6	30.7	35.0	39.3	43.8	48.4	53.1	57.9	66
67									3.6	7.3	11.1	15.0	18.9	23.0	27.2	31.5	35.8	40.3	44.8	49.5	54.3	67
68										3.7	7.5	11.4	15.3	19.4	23.6	27.9	32.2	36.7	41.2	45.9	50.7	68
69											3.8	7.7	11.6	15.7	19.8	24.1	28.4	32.9	37.5	42.2	47.0	69
70												3.9	7.8	11.9	16.1	20.3	24.7	29.2	33.7	38.4	43.2	70
71													4.0	8.1	12.2	16.5	20.8	25.3	29.8	34.5	39.3	71
72														4.1	8.2	12.5	16.8	21.3	25.9	30.6	35.4	72
73															4.1	8.4	12.7	17.2	21.8	26.5	31.3	73
74																4.3	8.6	13.1	17.6	22.3	27.1	74
75																	4.3	8.8	13.4	18.0	22.9	75
76																		4.5	9.0	13.7	18.5	76
77																			4.5	9.2	14.0	77
78																				4.7	9.5	78
79																					4.8	79

PERMISSION KAHOE AIR BALANCE COMPANY

VELOCITY OR CFM CORRECTION FOR TEMPERATURE AT 29.92 BAROMETRIC PRESSURE

DB Temperature °F	Density Lbs./Cu.Ft.	Relative Density	CFM Correction Factor
-25	.09134	1.219	.906
-20	.09028	1.205	.911
-15	.08926	1.191	.916
-10	.08827	1.178	.922
-5	.08732	1.165	.927
0	.08635	1.152	.932
5	.08542	1.140	.937
10	.08452	1.128	.942
15	.08363	1.116	.947
20	.08275	1.104	.952
22	.08241	1.100	.954
24	.08207	1.095	.956
26	.08173	1.091	.958
28	.08140	1.086	.960
30	.08107	1.082	.962
32	.08074	1.077	.963
34	.08042	1.073	.965
36	.08009	1.069	.967
38	.07978	1.064	.969
40	.07945	1.060	.971
42	.07914	1.056	.973
44	.07880	1.051	.975
46	.07850	1.047	.977
48	.07819	1.043	.979
50	.07788	1.039	.981
52	.07758	1.035	.983
54	.07728	1.031	.985
56	.07698	1.027	.987

PERMISSION KAHOE AIR BALANCE COMPANY

VELOCITY OR CFM CORRECTION FOR TEMPERATURE AT 29.92 BAROMETRIC PRESSURE

DB Temperature °F	Density Lbs./Cu.Ft.	Relative Density	CFM Correction Factor
58	.07668	1.023	.989
60	.07640	1.019	.990
62	.07610	1.015	.992
64	.07580	1.011	.994
66	.07552	1.008	.996
68	.07524	1.004	.998
70	.07495	1.000	1.000
72	.07468	.996	1.002
74	.07440	.993	1.004
76	.07412	.989	1.006
78	.07384	.985	1.007
80	.07357	.982	1.009
82	.07330	.978	1.011
84	.07302	.974	1.013
86	.07275	.971	1.015
88	.07248	.967	1.017
90	.07223	.964	1.019
92	.07196	.960	1.021
94	.07170	.957	1.022
96	.07144	.953	1.024
98	.07120	.950	1.026
100	.07093	.946	1.028
102	.07067	.943	1.030
104	.07042	.940	1.032
106	.07017	.936	1.034
108	.06993	.933	1.035
110	.06968	.930	1.037
115	.06908	.922	1.042

PERMISSION KAHOE AIR BALANCE COMPANY

VELOCITY OR CFM CORRECTION
TEMPERATURE AT 29.92 BAROMETRIC PRESSURE

DB Temperature °F	Density Lbs./Cu.Ft.	Relative Density	CFM Correction Factor
120	.06849	.914	1.046
125	.06790	.906	1.051
130	.06732	.898	1.055
135	.06676	.891	1.060
140	.06614	.882	1.065
145	.06565	.876	1.068
150	.06512	.869	1.072
155	.06459	.862	1.077
160	.06406	.855	1.082
165	.06356	.848	1.086
170	.06303	.841	1.090
175	.06258	.835	1.094
180	.06206	.828	1.099
185	.06161	.822	1.103
190	.06108	.815	1.108
195	.06064	.809	1.112
200	.06019	.803	1.116
210	.05929	.791	1.124
220	.05839	.779	1.133
230	.05756	.768	1.141
240	.05674	.757	1.149
250	.05600	.747	1.157
260	.05516	.736	1.166
270	.05441	.726	1.174
280	.05366	.716	1.182
290	.05299	.707	1.189
300	.05224	.697	1.198
310	.05157	.688	1.206

PERMISSION KAHOE AIR BALANCE COMPANY

VELOCITY OR CFM CORRECTION FOR TEMPERATURE AT 29.92 BAROMETRIC PRESSURE

DB Temperature °F	Density Lbs./Cu.Ft.	Relative Density	CFM Correction Factor
320	.05097	.680	1.213
330	.05029	.671	1.221
340	.04962	.662	1.229
350	.04902	.654	1.237
360	.04842	.646	1.244
370	.04782	.638	1.252
380	.04729	.631	1.259
390	.04677	.624	1.266
400	.04617	.616	1.274
425	.04490	.599	1.292
450	.04362	.582	1.311
475	.04250	.567	1.328
500	.04137	.552	1.346
525	.04032	.538	1.363
550	.03935	.525	1.380
575	.03837	.512	1.398
600	.03748	.500	1.414
625	.03658	.488	1.431
650	.03575	.477	1.448
675	.03500	.467	1.463
700	.03425	.457	1.479
725	.03350	.447	1.496
750	.03283	.438	1.511
775	.03215	.429	1.527
800	.03155	.421	1.541
825	.03088	.412	1.558
850	.03028	.404	1.573
875	.02976	.397	1.587

PERMISSION KAHOE AIR BALANCE COMPANY

VELOCITY OR CFM CORRECTION FOR ALTITUDE

ALTITUDE IN FEET	BAROMETRIC PRESSURE Inches Mercury	BAROMETRIC PRESSURE Lbs./Sq.In.	SPECIFIC VOLUME Cu.Ft./Lb.	AIR DENSITY	RELATIVE DENSITY	CFM TRANSMISSION FACTOR	CFM CORRECTION FACTOR
0	29.92	14.7	13.34	.0750	1.000	1.080	1.000
100	29.81	14.64	13.389	.0747	.996	1.076	1.002
200	29.70	14.58	13.439	.0745	.993	1.073	1.003
300	29.60	14.82	13.488	.0742	.989	1.068	1.005
400	29.49	14.46	13.538	.0739	.985	1.064	1.007
500	29.38	14.40	13.587	.0736	.981	1.060	1.010
600	29.28	14.36	13.636	.0734	.978	1.057	1.011
700	29.17	14.32	13.686	.0731	.975	1.053	1.013
800	29.06	14.28	13.735	.0728	.971	1.048	1.015
900	28.96	14.24	13.785	.0725	.967	1.044	1.017
1000	28.85	14.20	13.834	.0723	.964	1.041	1.019
1100	28.75	14.14	13.883	.0720	.960	1.037	1.021
1200	28.65	14.08	13.933	.0718	.957	1.034	1.022
1300	28.54	14.02	13.982	.0716	.954	1.031	1.024
1400	28.44	13.96	14.031	.0713	.951	1.027	1.026
1500	28.33	13.90	14.081	.0710	.947	1.022	1.028
1600	28.23	13.86	14.130	.0708	.944	1.020	1.029
1700	28.13	13.82	14.179	.0705	.940	1.015	1.031
1800	28.02	13.78	14.228	.0702	.936	1.011	1.034

PERMISSION KAHOE AIR BALANCE COMPANY

VELOCITY OR CFM CORRECTION FOR ALTITUDE

ALTITUDE IN FEET	BAROMETRIC PRESSURE Inches Mercury	BAROMETRIC PRESSURE Lbs./Sq.In.	SPECIFIC VOLUME Cu.Ft./Lb.	AIR DENSITY	RELATIVE DENSITY	CFM TRANSMISSION FACTOR	CFM CORRECTION FACTOR
1900	27.92	13.74	14.278	.0700	.933	1.008	1.035
2000	27.82	13.70	14.327	.0698	.930	1.005	1.037
2100	27.72	13.64	14.363	.0695	.926	1.001	1.039
2200	27.62	13.58	14.399	.0692	.923	.995	1.041
2300	27.52	13.52	14.435	.0690	.920	.994	1.043
2400	27.41	13.46	14.471	.0687	.916	.989	1.045
2500	27.31	13.40	14.507	.0685	.913	.986	1.046
2600	27.21	13.36	14.543	.0682	.909	.982	1.049
2700	27.11	13.32	14.579	.0680	.906	.979	1.050
2800	27.01	13.28	14.615	.0677	.903	.975	1.053
2900	26.91	13.24	14.651	.0675	.900	.972	1.054
3000	26.81	13.20	14.687	.0672	.896	.968	1.056
3200	26.61	13.10	14.836	.0667	.889	.960	1.060
3400	26.42	13.00	14.986	.0662	.883	.953	1.064
3600	26.23	12.90	15.135	.0658	.877	.948	1.068
3800	26.03	12.80	15.285	.0653	.870	.940	1.072
4000	25.84	12.70	15.434	.0648	.864	.933	1.076
4200	25.65	12.60	15.554	.0644	.858	.927	1.079
4400	25.46	12.50	15.674	.0638	.851	.919	1.084

PERMISSION KAHOE AIR BALANCE COMPANY

VELOCITY OR CFM CORRECTION FOR ALTITUDE

ALTITUDE IN FEET	BAROMETRIC PRESSURE Inches Mercury	PRESSURE Lbs./Sq.In.	SPECIFIC VOLUME Cu.Ft./Lb.	AIR DENSITY	RELATIVE DENSITY	CFM TRANSMISSION FACTOR	CFM CORRECTION FACTOR
4600	25.27	12.40	15.795	.0634	.845	.913	1.088
4800	25.08	12.30	15.915	.0629	.839	.906	1.092
5000	24.89	12.20	16.035	.0624	.832	.899	1.096
5200	24.71	12.12	16.167	.0619	.825	.891	1.101
5400	24.52	12.04	16.299	.0614	.819	.884	1.105
5600	24.34	11.96	16.431	.0610	.813	.878	1.109
5800	24.16	11.88	16.563	.0605	.807	.871	1.113
6000	23.98	11.80	16.695	.0599	.799	.863	1.119
6200	23.80	11.70	16.803	.0596	.794	.858	1.122
6400	23.62	11.60	16.911	.0592	.789	.852	1.126
6600	23.45	11.50	17.018	.0588	.784	.847	1.129
6800	23.27	11.40	17.126	.0584	.779	.841	1.133
7000	23.09	11.30	17.234	.0581	.774	.835	1.136
7200	22.90	11.22	17.397	.0575	.767	.828	1.142
7400	22.70	11.14	17.560	.0570	.760	.821	1.147
7600	22.51	11.06	17.724	.0565	.753	.814	1.152
7800	22.32	10.98	17.887	.0560	.746	.806	1.157
8000	22.12	10.90	18.050	.0554	.739	.798	1.163
8200	21.97	10.82	18.171	.0551	.734	.793	1.167

PERMISSION KAHOE AIR BALANCE COMPANY

A.27
Appendix

VELOCITY OR CFM CORRECTION FOR ALTITUDE

ALTITUDE IN FEET	BAROMETRIC PRESSURE Inches Mercury	BAROMETRIC PRESSURE Lbs./Sq.In.	SPECIFIC VOLUME Cu.Ft./Lb.	AIR DENSITY	RELATIVE DENSITY	CFM TRANSMISSION FACTOR	CFM CORRECTION FACTOR
8400	21.82	10.74	18.293	.0547	.729	.788	1.171
8600	21.68	10.66	18.414	.0544	.725	.783	1.174
8800	21.53	10.58	18.536	.0540	.720	.778	1.179
9000	21.38	10.50	18.657	.0536	.715	.772	1.183
9200	21.22	10.43	18.809	.0532	.709	.766	1.187
9400	21.06	10.34	18.961	.0528	.704	.760	1.192
9600	20.89	10.22	19.114	.0524	.698	.755	1.196
9800	20.73	10.18	19.265	.0520	.893	.749	1.201
10000	20.57	10.10	19.418	.0515	.687	.742	1.207
10200	20.42	10.03	19.564	.0512	.682	.737	1.210
10400	20.27	9.96	19.711	.0508	.677	.731	1.215
10600	20.13	9.88	19.857	.0504	.672	.726	1.220
10800	19.98	9.81	20.004	.0500	.667	.720	1.225
11000	19.83	9.74	20.150	.0497	.662	.715	1.228
11200	19.68	9.67	20.298	.0493	.657	.710	1.233
11400	19.54	9.60	20.445	.0490	.653	.705	1.237
11600	19.39	9.52	20.593	.0486	.648	.700	1.242
11800	19.24	9.45	20.740	.0482	.643	.694	1.247
12000	19.09	9.38	20.888	.0479	.638	.689	1.251

PERMISSION KAHOE AIR BALANCE COMPANY

VELOCITY OR CFM CORRECTION FOR ALTITUDE

ALTITUDE IN FEET	BAROMETRIC PRESSURE Inches Mercury	BAROMETRIC PRESSURE Lbs./Sq.In.	SPECIFIC VOLUME Cu.Ft./Lb.	AIR DENSITY	RELATIVE DENSITY	CFM TRANSMISSION FACTOR	CFM CORRECTION FACTOR
12200	18.95	9.31	21.062	.0475	.633	.684	1.257
12400	18.80	9.24	21.236	.0471	.628	.678	1.262
12600	18.65	9.16	21.411	.0467	.623	.673	1.267
12800	18.50	9.09	21.585	.0464	.618	.667	1.271
13000	18.36	9.02	21.759	.0460	.613	.662	1.277
13200	18.21	8.95	21.928	.0456	.608	.657	1.282
13400	18.06	8.88	22.097	.0452	.603	.651	1.288
13600	17.91	8.80	22.267	.0449	.598	.646	1.292
13800	17.77	8.73	22.436	.0446	.594	.642	1.299
14000	17.62	8.66	22.605	.0442	.589	.636	1.303
14200	17.47	8.59	22.814	.0438	.584	.631	1.309
14400	17.32	8.52	23.024	.0434	.579	.625	1.315
14600	17.18	8.44	23.233	.0431	.574	.620	1.319
14800	17.03	8.37	23.443	.0427	.569	.615	1.325
15000	16.88	8.30	23.652	.0423	.564	.609	1.332

PERMISSION KAHOE AIR BALANCE COMPANY

DENSITY - 1
WEIGHT OF SATURATED AND PARTLY SATURATED AIR

Dry Bulb Temp. °F t	Barometric Pressure in Inches of Mercury				Increase in weight per .01" rise in barometer	Density increase per degree W.B. depression
	20	20.5	21	21.5		
30	.05401	.05537	.05672	.05808	.00027	.000017
31	.05389	.05524	.05660	.05795	.00027	.000017
32	.05377	.05512	.05647	.05782	.00027	.000017
33	.05366	.05500	.05635	.05770	.00027	.000018
34	.05354	.05488	.05623	.05757	.00027	.000018
35	.05342	.05476	.05611	.05745	.00027	.000018
36	.05331	.05465	.05598	.05732	.00027	.000018
37	.05319	.05453	.05586	.05720	.00027	.000019
38	.05308	.05441	.05574	.05707	.00027	.000019
39	.05296	.05429	.05562	.05695	.00027	.000019
40	.05285	.05417	.05550	.05683	.00027	.000019
41	.05273	.05406	.05538	.05670	.00026	.000020
42	.05261	.05394	.05526	.05658	.00026	.000020
43	.05250	.05382	.05514	.05646	.00026	.000020
44	.05238	.05370	.05502	.05633	.00026	.000020
45	.05227	.05358	.05490	.05621	.00026	.000020
46	.05215	.05346	.05478	.05609	.00026	.000021
47	.05204	.05335	.05466	.05597	.00026	.000021
48	.05192	.05323	.05454	.05584	.00026	.000021
49	.05181	.05311	.05442	.05572	.00026	.000022
50	.05170	.05300	.05430	.05560	.00026	.000022
51	.05158	.05288	.05418	.05548	.00026	.000022
52	.05146	.05276	.05406	.05535	.00026	.000023
53	.05135	.05264	.05394	.05523	.00026	.000023
54	.05123	.05252	.05382	.05511	.00026	.000023
55	.05112	.05241	.05370	.05498	.00026	.000024

PERMISSION KAHOE AIR BALANCE COMPANY

DENSITY - 1

WEIGHT OF SATURATED AND PARTLY SATURATED AIR

Dry Bulb Temp. °F t	Barometric Pressure in Inches of Mercury				Increase in weight per .01" rise in barometer	Density increase per degree W.B. depression
	22	22.5	23	23.5		
30	.05943	.06078	.06214	.06349	.00027	.000017
31	.05930	.06065	.06200	.06336	.00027	.000017
32	.05917	.06052	.06187	.06322	.00027	.000017
33	.05904	.06039	.06174	.06308	.00027	.000018
34	.05891	.06026	.06160	.06295	.00027	.000018
35	.05879	.06013	.06147	.06281	.00027	.000018
36	.05866	.06000	.06134	.06267	.00027	.000018
37	.05853	.05987	.06120	.06254	.00027	.000019
38	.05841	.05974	.06107	.06241	.00027	.000019
39	.05828	.05961	.06094	.06227	.00027	.000019
40	.05816	.05948	.06081	.06214	.00027	.000019
41	.05803	.05935	.06068	.06200	.00026	.000020
42	.05790	.05922	.06055	.06187	.00026	.000020
43	.05778	.05910	.06042	.06173	.00026	.000020
44	.05765	.05897	.06028	.06160	.00026	.000020
45	.05752	.05884	.06015	.06147	.00026	.000020
46	.05740	.05871	.06002	.06133	.00026	.000021
47	.05727	.05858	.05989	.06120	.00026	.000021
48	.05715	.05846	.05976	.06107	.00026	.000021
49	.05703	.05833	.05963	.06094	.00026	.000022
50	.05690	.05821	.05951	.06081	.00026	.000022
51	.05678	.05808	.05937	.06067	.00026	.000022
52	.05665	.05795	.05924	.06054	.00026	.000023
53	.05652	.05782	.05911	.06041	.00026	.000023
54	.05640	.05769	.05898	.06027	.00026	.000023
55	.05627	.05756	.05885	.06014	.00026	.000024

PERMISSION KAHOE AIR BALANCE COMPANY

DENSITY - 1

WEIGHT OF SATURATED AND PARTLY SATURATED AIR

Dry Bulb Temp. °F t	Barometric Pressure in Inches of Mercury				Increase in weight per .01" rise in barometer	Density increase per degree W.B. depression
	24	24.5	25	25.5		
30	.06485	.06620	.06756	.06891	.00027	.000017
31	.06471	.06606	.06741	.06876	.00027	.000017
32	.06457	.06592	.06727	.06862	.00027	.000017
33	.06443	.06577	.06712	.06847	.00027	.000018
34	.06429	.06563	.06698	.06832	.00027	.000018
35	.06415	.06549	.06683	.06817	.00027	.000018
36	.06401	.06535	.06669	.06803	.00027	.000018
37	.06388	.06521	.06655	.06788	.00027	.000019
38	.06374	.06507	.06640	.06774	.00027	.000019
39	.06360	.06493	.06626	.06759	.00027	.000019
40	.06347	.06479	.06612	.06745	.00027	.000019
41	.06333	.06465	.06598	.06730	.00026	.000020
42	.06319	.06451	.06584	.06716	.00026	.000020
43	.06305	.06437	.06569	.06701	.00026	.000020
44	.06292	.06423	.06555	.06687	.00026	.000020
45	.06278	.06410	.06541	.06672	.00026	.000020
46	.06265	.06396	.06527	.06658	.00026	.000021
47	.06251	.06382	.06513	.06644	.00026	.000021
48	.06238	.06368	.06499	.06630	.00026	.000021
49	.06224	.06355	.06485	.06616	.00026	.000022
50	.06211	.06341	.06471	.06601	.00026	.000022
51	.06197	.06327	.06457	.06587	.00026	.000022
52	.06184	.06313	.06443	.06572	.00026	.000023
53	.06170	.06299	.06429	.06558	.00026	.000023
54	.06156	.06286	.06415	.06544	.00026	.000023
55	.06143	.06272	.06401	.06530	.00026	.000024

PERMISSION KAHOE AIR BALANCE COMPANY

DENSITY - 1

WEIGHT OF SATURATED AND PARTLY SATURATED AIR

Dry Bulb Temp. °F t	Barometric Pressure in Inches of Mercury				Increase in weight per .01" rise in barometer	Density increase per degree W.B. depression
	26	26.5	27	27.5		
30	.07027	.07162	.07298	.07433	.00027	.000017
31	.07012	.07147	.07282	.07417	.00027	.000017
32	.06996	.07131	.07266	.07401	.00027	.000017
33	.06981	.07116	.07251	.07385	.00027	.000018
34	.06966	.07101	.07235	.07369	.00027	.000018
35	.06951	.07086	.07220	.07354	.00027	.000018
36	.06937	.07070	.07204	.07338	.00027	.000018
37	.06922	.07055	.07189	.07322	.00027	.000019
38	.06907	.07040	.07174	.07307	.00027	.000019
39	.06892	.07025	.07158	.07291	.00027	.000019
40	.06878	.07011	.07143	.07276	.00027	.000019
41	.06863	.06995	.07128	.07260	.00026	.000020
42	.06848	.06980	.07112	.07245	.00026	.000020
43	.06833	.06965	.07097	.07229	.00026	.000020
44	.06819	.06950	.07082	.07214	.00026	.000020
45	.06804	.06935	.07067	.07198	.00026	.000020
46	.06789	.06921	.07052	.07183	.00026	.000021
47	.06775	.06906	.07037	.07168	.00026	.000021
48	.06760	.06891	.07022	.07152	.00026	.000021
49	.06746	.06876	.07007	.07137	.00026	.000022
50	.06732	.06862	.06992	.07122	.00026	.000022
51	.06717	.06847	.06977	.07106	.00026	.000022
52	.06702	.06832	.06961	.07091	.00026	.000023
53	.06688	.06817	.06946	.07076	.00026	.000023
54	.06673	.06802	.06931	.07060	.00026	.000023
55	.06658	.06787	.06916	.07045	.00026	.000024

PERMISSION KAHOE AIR BALANCE COMPANY

DENSITY - 1

WEIGHT OF SATURATED AND PARTLY SATURATED AIR

Dry Bulb Temp. °F t	Barometric Pressure in Inches of Mercury				Increase in weight per .01" rise in barometer	Density increase per degree W.B. depression
	28	28.5	29	29.5		
30	.07569	.07704	.07839	.07975	.00027	.000017
31	.07552	.07687	.07823	.07958	.00027	.000017
32	.07536	.07671	.07806	.07941	.00027	.000017
33	.07520	.07655	.07789	.07924	.00027	.000018
34	.07504	.07638	.07773	.07907	.00027	.000018
35	.07488	.07622	.07756	.07890	.00027	.000018
36	.07472	.07606	.07740	.07873	.00027	.000018
37	.07456	.07590	.07723	.07857	.00027	.000019
38	.07440	.07573	.07707	.07840	.00027	.000019
39	.07424	.07557	.07690	.07824	.00027	.000019
40	.07409	.07542	.07674	.07807	.00027	.000019
41	.07393	.07525	.07658	.07790	.00026	.000020
42	.07377	.07509	.07641	.07774	.00026	.000020
43	.07361	.07493	.07625	.07757	.00026	.000020
44	.07345	.07477	.07609	.07740	.00026	.000020
45	.07330	.07461	.07593	.07724	.00026	.000020
46	.07314	.07445	.07576	.07708	.00026	.000021
47	.07298	.07429	.07560	.07691	.00026	.000021
48	.07283	.07414	.07544	.07675	.00026	.000021
49	.07268	.07398	.07528	.07659	.00026	.000022
50	.07252	.07382	.07512	.07643	.00026	.000022
51	.07236	.07366	.07496	.07626	.00026	.000022
52	.07221	.07350	.07480	.07610	.00026	.000023
53	.07205	.07334	.07464	.07593	.00026	.000023
54	.07189	.07319	.07448	.07577	.00026	.000023
55	.07174	.07303	.07432	.07561	.00026	.000024

PERMISSION KAHOE AIR BALANCE COMPANY

DENSITY - 1

WEIGHT OF SATURATED AND PARTLY SATURATED AIR

Dry Bulb Temp. °F t	Barometric Pressure in Inches of Mercury			Increase in weight per .01" rise in barometer	Density increase per degree W.B. depression
	30	30.5	31		
30	.08110	.08246	.08381	.00027	.000017
31	.08093	.08228	.08363	.00027	.000017
32	.08076	.08211	.08346	.00027	.000017
33	.08058	.08193	.08328	.00027	.000018
34	.08041	.08176	.08310	.00027	.000018
35	.08024	.08158	.08292	.00027	.000018
36	.08007	.08141	.08275	.00027	.000018
37	.07990	.08124	.08257	.00027	.000019
38	.07973	.08107	.08240	.00027	.000019
39	.07957	.08090	.08223	.00027	.000019
40	.07940	.08073	.08205	.00027	.000019
41	.07923	.08055	.08188	.00026	.000020
42	.07906	.08038	.08170	.00026	.000020
43	.07889	.08021	.08153	.00026	.000020
44	.07872	.08004	.08136	.00026	.000020
45	.07855	.07987	.08118	.00026	.000020
46	.07839	.07970	.08101	.00026	.000021
47	.07822	.07953	.08084	.00026	.000021
48	.07806	.07936	.08067	.00026	.000021
49	.07789	.07920	.08050	.00026	.000022
50	.07773	.07903	.08033	.00026	.000022
51	.07756	.07886	.08016	.00026	.000022
52	.07739	.07869	.07999	.00026	.000023
53	.07723	.07852	.07981	.00026	.000023
54	.07706	.07835	.07964	.00026	.000023
55	.07690	.07818	.07947	.00026	.000024

PERMISSION KAHOE AIR BALANCE COMPANY

DENSITY - 1
WEIGHT OF SATURATED AND PARTLY SATURATED AIR

Dry Bulb Temp. °F t	Barometric Pressure in Inches of Mercury				Increase in weight per .01" rise in barometer	Density increase per degree W.B. depression
	20	20.5	21	21.5		
56	.05100	.05229	.05358	.05486	.00026	.000024
57	.05089	.05217	.05346	.05474	.00026	.000025
58	.05078	.05206	.05334	.05462	.00026	.000025
59	.05066	.05194	.05322	.05450	.00026	.000025
60	.05055	.05183	.05310	.05438	.00026	.000026
61	.05043	.05171	.05298	.05425	.00026	.000026
62	.05031	.05159	.05286	.05413	.00026	.000027
63	.05020	.05147	.05274	.05400	.00026	.000027
64	.05008	.05135	.05261	.05388	.00026	.000028
65	.04996	.05123	.05249	.05376	.00026	.000028
66	.04985	.05111	.05237	.05363	.00026	.000029
67	.04973	.05099	.05225	.05351	.00026	.000029
68	.04962	.05088	.05213	.05339	.00026	.000030
69	.04950	.05076	.05201	.05327	.00026	.000030
70	.04939	.05064	.05189	.05315	.00026	.000031
71	.04927	.05052	.05177	.05302	.00025	.000031
72	.04915	.05040	.05164	.05289	.00025	.000032
73	.04903	.05027	.05152	.05277	.00025	.000033
74	.04891	.05015	.05140	.05264	.00025	.000033
75	.04879	.05003	.05127	.05251	.00025	.000034
76	.04867	.04991	.05115	.05239	.00025	.000034
77	.04855	.04979	.05103	.05226	.00025	.000035
78	.04844	.04967	.05090	.05214	.00025	.000036
79	.04832	.04955	.05078	.05201	.00025	.000036
80	.04820	.04943	.05066	.05189	.00025	.000037
81	.04808	.04931	.05053	.05176	.00025	.000038

PERMISSION KAHOE AIR BALANCE COMPANY

DENSITY - 1
WEIGHT OF SATURATED AND PARTLY SATURATED AIR

Dry Bulb Temp. °F t	Barometric Pressure in Inches of Mercury				Increase in weight per .01" rise in barometer	Density increase per degree W.B. depression
	22	22.5	23	23.5		
56	.05615	.05744	.05872	.06001	.00026	.000024
57	.05602	.05731	.05859	.05988	.00026	.000025
58	.05590	.05718	.05846	.05975	.00026	.000025
59	.05578	.05706	.05834	.05961	.00026	.000025
60	.05566	.05693	.05821	.05948	.00026	.000026
61	.05553	.05680	.05808	.05935	.00026	.000026
62	.05540	.05667	.05794	.05922	.00026	.000027
63	.05527	.05654	.05781	.05908	.00026	.000027
64	.05515	.05641	.05768	.05895	.00026	.000028
65	.05502	.05629	.05755	.05881	.00026	.000028
66	.05490	.05616	.05742	.05868	.00026	.000029
67	.05477	.05603	.05729	.05855	.00026	.000029
68	.05465	.05590	.05716	.05842	.00026	.000030
69	.05452	.05578	.05703	.05829	.00026	.000030
70	.05440	.05565	.05690	.05816	.00026	.000031
71	.05427	.05552	.05677	.05802	.00025	.000031
72	.05414	.05539	.05663	.05788	.00025	.000032
73	.05401	.05526	.05650	.05775	.00025	.000033
74	.05388	.05512	.05637	.05761	.00025	.000033
75	.05375	.05499	.05624	.05748	.00025	.000034
76	.05363	.05486	.05610	.05734	.00025	.000034
77	.05350	.05473	.05597	.05721	.00025	.000035
78	.05337	.05461	.05584	.05707	.00025	.000036
79	.05325	.05448	.05571	.05694	.00025	.000036
80	.05312	.05435	.05558	.05681	.00025	.000037
81	.05299	.05421	.05544	.05667	.00025	.000038

PERMISSION KAHOE AIR BALANCE COMPANY

DENSITY – 1
WEIGHT OF SATURATED AND PARTLY SATURATED AIR

Dry Bulb Temp. °F t	Barometric Pressure in Inches of Mercury				Increase in weight per .01" rise in barometer	Density increase per degree W.B. depression
	24	24.5	25	25.5		
56	.06129	.06258	.06387	.06515	.00026	.000024
57	.06116	.06244	.06373	.06501	.00026	.000025
58	.06103	.06231	.06359	.06487	.00026	.000025
59	.06089	.06217	.06345	.06473	.00026	.000025
60	.06076	.06204	.06331	.06459	.00026	.000026
61	.06062	.06190	.06317	.06445	.00026	.000026
62	.06049	.06176	.06303	.06430	.00026	.000027
63	.06035	.06162	.06289	.06416	.00026	.000027
64	.06021	.06148	.06275	.06401	.00026	.000028
65	.06008	.06134	.06261	.06387	.00026	.000028
66	.05994	.06121	.06247	.06373	.00026	.000029
67	.05981	.06107	.06233	.06359	.00026	.000029
68	.05968	.06093	.06219	.06345	.00026	.000030
69	.05954	.06080	.06205	.06331	.00026	.000030
70	.05941	.06066	.06191	.06317	.00026	.000031
71	.05927	.06052	.06177	.06302	.00025	.000031
72	.05913	.06038	.06163	.06287	.00025	.000032
73	.05899	.06024	.06148	.06273	.00025	.000033
74	.05885	.06010	.06134	.06258	.00025	.000033
75	.05872	.05996	.06120	.06244	.00025	.000034
76	.05858	.05982	.06106	.06229	.00025	.000034
77	.05844	.05968	.06092	.06215	.00025	.000035
78	.05831	.05954	.06077	.06201	.00025	.000036
79	.05817	.05940	.06063	.06187	.00025	.000036
80	.05804	.05927	.06050	.06172	.00025	.000037
81	.05789	.05912	.06035	.06157	.00025	.000038

PERMISSION KAHOE AIR BALANCE COMPANY

DENSITY - 1
WEIGHT OF SATURATED AND PARTLY SATURATED AIR

Dry Bulb Temp. °F t	Barometric Pressure in Inches of Mercury				Increase in weight per .01" rise in barometer	Density increase per degree W.B. depression
	26	26.5	27	27.5		
56	.06644	.06773	.06901	.07030	.00026	.000024
57	.06630	.06758	.06886	.07015	.00026	.000025
58	.06615	.06743	.06872	.07000	.00026	.000025
59	.06601	.06729	.06857	.06985	.00026	.000025
60	.06587	.06714	.06842	.06970	.00026	.000026
61	.06572	.06699	.06827	.06954	.00026	.000026
62	.06557	.06684	.06812	.06939	.00026	.000027
63	.06543	.06670	.06796	.06923	.00026	.000027
64	.06528	.06655	.06781	.06908	.00026	.000028
65	.06514	.06640	.06766	.06893	.00026	.000028
66	.06499	.06625	.06751	.06878	.00026	.000029
67	.06485	.06611	.06737	.06863	.00026	.000029
68	.06470	.06596	.06722	.06847	.00026	.000030
69	.06456	.06582	.06707	.06832	.00026	.000030
70	.06442	.06567	.06692	.06818	.00026	.000031
71	.06427	.06552	.06677	.06802	.00025	.000031
72	.06412	.06537	.06662	.06786	.00025	.000032
73	.06397	.06522	.06646	.06771	.00025	.000033
74	.06383	.06507	.06631	.06755	.00025	.000033
75	.06368	.06492	.06616	.06740	.00025	.000034
76	.06353	.06477	.06601	.06725	.00025	.000034
77	.06339	.06462	.06586	.06710	.00025	.000035
78	.06324	.06448	.06571	.06694	.00025	.000036
79	.06310	.06433	.06556	.06679	.00025	.000036
80	.06295	.06418	.06541	.06664	.00025	.000037
81	.06280	.06403	.06525	.06648	.00025	.000038

PERMISSION KAHOE AIR BALANCE COMPANY

DENSITY - 1
WEIGHT OF SATURATED AND PARTLY SATURATED AIR

Dry Bulb Temp. °F t	Barometric Pressure in Inches of Mercury				Increase in weight per .01" rise in barometer	Density increase per degree W.B. depression
	28	28.5	29	29.5		
56	.07159	.07287	.07416	.07544	.00026	.000024
57	.07143	.07272	.07400	.07528	.00026	.000025
58	.07128	.07256	.07384	.07512	.00026	.000025
59	.07113	.07240	.07368	.07496	.00026	.000025
60	.07097	.07225	.07353	.07480	.00026	.000026
61	.07082	.07209	.07336	.07464	.00026	.000026
62	.07066	.07193	.07320	.07447	.00026	.000027
63	.07050	.07177	.07304	.07431	.00026	.000027
64	.07035	.07161	.07288	.07415	.00026	.000028
65	.07019	.07146	.07272	.07399	.00026	.000028
66	.07004	.07130	.07256	.07382	.00026	.000029
67	.06989	.07114	.07240	.07366	.00026	.000029
68	.06973	.07099	.07225	.07350	.00026	.000030
69	.06958	.07083	.07209	.07334	.00026	.000030
70	.06943	.07068	.07193	.07318	.00026	.000031
71	.06927	.07052	.07177	.07302	.00025	.000031
72	.06911	.07036	.07161	.07285	.00025	.000032
73	.06895	.07020	.07144	.07269	.00025	.000033
74	.06880	.07004	.07128	.07253	.00025	.000033
75	.06864	.06988	.07112	.07236	.00025	.000034
76	.06849	.06972	.07096	.07220	.00025	.000034
77	.06833	.06957	.07080	.07204	.00025	.000035
78	.06818	.06941	.07064	.07188	.00025	.000036
79	.06802	.06925	.07049	.07172	.00025	.000036
80	.06787	.06910	.07033	.07156	.00025	.000037
81	.06771	.06894	.07016	.07139	.00025	.000038

PERMISSION KAHOE AIR BALANCE COMPANY

DENSITY – 1
WEIGHT OF SATURATED AND PARTLY SATURATED AIR

Dry Bulb Temp. °F t	Barometric Pressure in Inches of Mercury			Increase in weight per .01" rise in barometer	Density increase per degree W.B. depression
	30	30.5	31		
56	.07673	.07802	.07930	.00026	.000024
57	.07657	.07785	.07913	.00026	.000025
58	.07640	.07768	.07897	.00026	.000025
59	.07624	.07752	.07880	.00026	.000025
60	.07608	.07736	.07863	.00026	.000026
61	.07591	.07719	.07846	.00026	.000026
62	.07575	.07702	.07829	.00026	.000027
63	.07558	.07685	.07812	.00026	.000027
64	.07541	.07668	.07795	.00026	.000028
65	.07525	.07651	.07778	.00026	.000028
66	.07509	.07635	.07761	.00026	.000029
67	.07492	.07618	.07744	.00026	.000029
68	.07476	.07602	.07727	.00026	.000030
69	.07460	.07585	.07711	.00026	.000030
70	.07444	.07569	.07694	.00026	.000031
71	.07427	.07552	.07677	.00025	.000031
72	.07410	.07535	.07660	.00025	.000032
73	.07394	.07518	.07643	.00025	.000033
74	.07377	.07501	.07626	.00025	.000033
75	.07360	.07484	.07609	.00025	.000034
76	.07344	.07468	.07592	.00025	.000034
77	.07328	.07451	.07575	.00025	.000035
78	.07311	.07435	.07558	.00025	.000036
79	.07295	.07418	.07541	.00025	.000036
80	.07279	.07402	.07525	.00025	.000037
81	.07262	.07384	.07507	.00025	.000038

PERMISSION KAHOE AIR BALANCE COMPANY

DENSITY – 1
WEIGHT OF SATURATED AND PARTLY SATURATED AIR

Dry Bulb Temp. °F t	Barometric Pressure in Inches of Mercury				Increase in weight per .01" rise in barometer	Density increase per degree W.B. depression
	20	20.5	21	21.5		
82	.04795	.04918	.05040	.05163	.00024	.000039
83	.04783	.04905	.05027	.05150	.00024	.000039
84	.04771	.04893	.05015	.05137	.00024	.000040
85	.04758	.04880	.05002	.05124	.00024	.000041
86	.04746	.04867	.04989	.05111	.00024	.000042
87	.04734	.04855	.04976	.05098	.00024	.000043
88	.04721	.04843	.04964	.05085	.00024	.000043
89	.04709	.04830	.04951	.05072	.00024	.000044
90	.04697	.04818	.04939	.05059	.00024	.000045
91	.04684	.04804	.04925	.05045	.00024	.000046
92	.04671	.04791	.04911	.05032	.00024	.000047
93	.04658	.04778	.04898	.05018	.00024	.000048
94	.04645	.04765	.04884	.05004	.00024	.000049
95	.04632	.04751	.04871	.04991	.00024	.000050
96	.04619	.04738	.04858	.04977	.00024	.000051
97	.04606	.04725	.04844	.04964	.00024	.000052
98	.04593	.04712	.04831	.04950	.00024	.000053
99	.04581	.04699	.04818	.04937	.00024	.000054
100	.04568	.04686	.04805	.04923	.00024	.000055
101	.04554	.04672	.04790	.04909	.00024	.000056
102	.04540	.04658	.04776	.04894	.00024	.000057
103	.04526	.04644	.04762	.04879	.00024	.000058
104	.04512	.04630	.04747	.04865	.00024	.000059
105	.04498	.04615	.04733	.04850	.00023	.000060
106	.04484	.04601	.04719	.04836	.00023	.000061
107	.04470	.04588	.04705	.04822	.00023	.000062

PERMISSION KAHOE AIR BALANCE COMPANY

DENSITY - 1
WEIGHT OF SATURATED AND PARTLY SATURATED AIR

Dry Bulb Temp. °F t	Barometric Pressure in Inches of Mercury				Increase in weight per .01" rise in barometer	Density increase per degree W.B. depression
	22	22.5	23	23.5		
82	.05285	.05408	.05530	.05653	.00024	.000039
83	.05272	.05394	.05516	.05639	.00024	.000039
84	.05259	.05381	.05503	.05625	.00024	.000040
85	.05245	.05367	.05489	.05611	.00024	.000041
86	.05232	.05354	.05475	.05597	.00024	.000042
87	.05219	.05340	.05462	.05583	.00024	.000043
88	.05206	.05327	.05448	.05569	.00024	.000043
89	.05193	.05314	.05435	.05556	.00024	.000044
90	.05180	.05301	.05421	.05542	.00024	.000045
91	.05166	.05286	.05407	.05527	.00024	.000046
92	.05152	.05272	.05392	.05513	.00024	.000047
93	.05138	.05258	.05378	.05498	.00024	.000048
94	.05124	.05244	.05364	.05484	.00024	.000049
95	.05110	.05230	.05349	.05469	.00024	.000050
96	.05096	.05216	.05335	.05455	.00024	.000051
97	.05083	.05202	.05321	.05440	.00024	.000052
98	.05069	.05188	.05307	.05426	.00024	.000053
99	.05055	.05174	.05293	.05412	.00024	.000054
100	.05042	.05160	.05279	.05397	.00024	.000055
101	.05027	.05145	.05264	.05382	.00024	.000056
102	.05012	.05130	.05248	.05366	.00024	.000057
103	.04997	.05115	.05233	.05351	.00024	.000058
104	.04983	.05100	.05218	.05336	.00024	.000059
105	.04968	.05085	.05203	.05320	.00023	.000060
106	.04953	.05071	.05188	.05305	.00023	.000061
107	.04939	.05056	.05173	.05290	.00023	.000062

PERMISSION KAHOE AIR BALANCE COMPANY

DENSITY - 1
WEIGHT OF SATURATED AND PARTLY SATURATED AIR

Dry Bulb Temp. °F t	Barometric Pressure in Inches of Mercury				Increase in weight per .01" rise in barometer	Density increase per degree W.B. depression
	24	24.5	25	25.5		
82	.05775	.05898	.06020	.06142	.00024	.000039
83	.05761	.05883	.06005	.06128	.00024	.000039
84	.05747	.05869	.05991	.06113	.00024	.000040
85	.05733	.05854	.05976	.06098	.00024	.000041
86	.05718	.05840	.05962	.06083	.00024	.000042
87	.05704	.05826	.05947	.06068	.00024	.000043
88	.05690	.05812	.05933	.06054	.00024	.000043
89	.05677	.05797	.05918	.06039	.00024	.000044
90	.05663	.05783	.05904	.06025	.00024	.000045
91	.05648	.05768	.05889	.06009	.00024	.000046
92	.05633	.05753	.05873	.05994	.00024	.000047
93	.05618	.05738	.05858	.05978	.00024	.000048
94	.05603	.05723	.05843	.05963	.00024	.000049
95	.05589	.05708	.05828	.05947	.00024	.000050
96	.05574	.05693	.05813	.05932	.00024	.000051
97	.05559	.05679	.05798	.05917	.00024	.000052
98	.05545	.05664	.05783	.05902	.00024	.000053
99	.05530	.05649	.05768	.05887	.00024	.000054
100	.05516	.05635	.05753	.05872	.00024	.000055
101	.05500	.05619	.05737	.05855	.00024	.000056
102	.05485	.05603	.05721	.05839	.00024	.000057
103	.05469	.05587	.05705	.05823	.00024	.000058
104	.05453	.05571	.05689	.05806	.00024	.000059
105	.05438	.05555	.05673	.05790	.00023	.000060
106	.05422	.05540	.05657	.05774	.00023	.000061
107	.05407	.05524	.05641	.05758	.00023	.000062

PERMISSION KAHOE AIR BALANCE COMPANY

DENSITY - 1

WEIGHT OF SATURATED AND PARTLY SATURATED AIR

Dry Bulb Temp. °F t	Barometric Pressure in Inches of Mercury				Increase in weight per .01" rise in barometer	Density increase per degree W.B. depression
	26	26.5	27	27.5		
82	.06265	.06387	.06510	.06632	.00024	.000039
83	.06250	.06372	.06494	.06616	.00024	.000039
84	.06235	.06357	.06479	.06601	.00024	.000040
85	.06220	.06341	.06463	.06585	.00024	.000041
86	.06205	.06326	.06448	.06569	.00024	.000042
87	.06190	.06311	.06432	.06554	.00024	.000043
88	.06175	.06296	.06417	.06538	.00024	.000043
89	.06160	.06281	.06402	.06523	.00024	.000044
90	.06145	.06266	.06387	.06507	.00024	.000045
91	.06130	.06250	.06371	.06491	.00024	.000046
92	.06114	.06234	.06354	.06475	.00024	.000047
93	.06098	.06218	.06338	.06458	.00024	.000048
94	.06083	.06202	.06322	.06442	.00024	.000049
95	.06067	.06187	.06306	.06426	.00024	.000050
96	.06052	.06171	.06290	.06410	.00024	.000051
97	.06036	.06155	.06274	.06394	.00024	.000052
98	.06021	.06140	.06259	.06378	.00024	.000053
99	.06005	.06124	.06243	.06362	.00024	.000054
100	.05990	.06109	.06227	.06346	.00024	.000055
101	.05973	.06092	.06210	.06328	.00024	.000056
102	.05957	.06075	.06193	.06311	.00024	.000057
103	.05940	.06058	.06176	.06294	.00024	.000058
104	.05924	.06042	.06159	.06277	.00024	.000059
105	.05908	.06025	.06143	.06260	.00023	.000060
106	.05891	.06009	.06126	.06243	.00023	.000061
107	.05875	.05992	.06109	.06226	.00023	.000062

PERMISSION KAHOE AIR BALANCE COMPANY

DENSITY - 1
WEIGHT OF SATURATED AND PARTLY SATURATED AIR

Dry Bulb Temp. °F t	Barometric Pressure in Inches of Mercury				Increase in weight per .01" rise in barometer	Density increase per degree W.B. depression
	28	28.5	29	29.5		
82	.06755	.06877	.07000	.07122	.00024	.000039
83	.06739	.06861	.06983	.07105	.00024	.000039
84	.06723	.06845	.06967	.07089	.00024	.000040
85	.06707	.06829	.06950	.07072	.00024	.000041
86	.06691	.06813	.06934	.07056	.00024	.000042
87	.06675	.06796	.06918	.07039	.00024	.000043
88	.06659	.06781	.06902	.07023	.00024	.000043
89	.06644	.06765	.06886	.07006	.00024	.000044
90	.06628	.06749	.06869	.06990	.00024	.000045
91	.06611	.06732	.06852	.06973	.00024	.000046
92	.06595	.06715	.06835	.06956	.00024	.000047
93	.06578	.06698	.06818	.06938	.00024	.000048
94	.06562	.06682	.06801	.06921	.00024	.000049
95	.06545	.06665	.06785	.06904	.00024	.000050
96	.06529	.06648	.06768	.06887	.00024	.000051
97	.06513	.06632	.06751	.06870	.00024	.000052
98	.06496	.06615	.06734	.06853	.00024	.000053
99	.06480	.06599	.06718	.06837	.00024	.000054
100	.06464	.06583	.06701	.06820	.00024	.000055
101	.06447	.06565	.06683	.06802	.00024	.000056
102	.06429	.06547	.06666	.06784	.00024	.000057
103	.06412	.06530	.06648	.06766	.00024	.000058
104	.06395	.06512	.06630	.06748	.00024	.000059
105	.06378	.06495	.06613	.06730	.00023	.000060
106	.06360	.06478	.06595	.06712	.00023	.000061
107	.06343	.06460	.06578	.06695	.00023	.000062

PERMISSION KAHOE AIR BALANCE COMPANY

DENSITY - 1
WEIGHT OF SATURATED AND PARTLY SATURATED AIR

Dry Bulb Temp. °F t	Barometric Pressure in Inches of Mercury			Increase in weight per .01" rise in barometer	Density increase per degree W.B. depression
	30	30.5	31		
82	.07245	.07367	.07490	.00024	.000039
83	.07228	.07350	.07472	.00024	.000039
84	.07211	.07333	.07455	.00024	.000040
85	.07194	.07316	.07438	.00024	.000041
86	.07177	.07299	.07420	.00024	.000042
87	.07161	.07282	.07403	.00024	.000043
88	.07144	.07265	.07386	.00024	.000043
89	.07127	.07248	.07369	.00024	.000044
90	.07111	.07231	.07352	.00024	.000045
91	.07093	.07214	.07334	.00024	.000046
92	.07076	.07196	.07316	.00024	.000047
93	.07058	.07178	.07298	.00024	.000048
94	.07041	.07161	.07281	.00024	.000049
95	.07024	.07143	.07263	.00024	.000050
96	.07007	.07126	.07245	.00024	.000051
97	.06989	.07109	.07228	.00024	.000052
98	.06972	.07091	.07210	.00024	.000053
99	.06955	.07074	.07193	.00024	.000054
100	.06938	.07057	.07175	.00024	.000055
101	.06920	.07038	.07157	.00024	.000056
102	.06902	.07020	.07138	.00024	.000057
103	.06884	.07002	.07119	.00024	.000058
104	.06866	.06983	.07101	.00024	.000059
105	.06848	.06965	.07082	.00023	.000060
106	.06830	.06947	.07064	.00023	.000061
107	.06812	.06929	.07046	.00023	.000062

PERMISSION KAHOE AIR BALANCE COMPANY

DENSITY - 1
WEIGHT OF SATURATED AND PARTLY SATURATED AIR

Dry Bulb Temp. °F t	Barometric Pressure in Inches of Mercury				Increase in weight per .01" rise in barometer	Density increase per degree W.B. depression
	20	20.5	21	21.5		
108	.04457	.04574	.04690	.04807	.00023	.000063
109	.04443	.04560	.04676	.04793	.00023	.000064
110	.04429	.04546	.04662	.04779	.00023	.000065

PERMISSION KAHOE AIR BALANCE COMPANY

DENSITY - 1
WEIGHT OF SATURATED AND PARTLY SATURATED AIR

Dry Bulb Temp. °F t	Barometric Pressure in Inches of Mercury				Increase in weight per .01" rise in barometer	Density increase per degree W.B. depression
	26	26.5	27	27.5		
108	.05859	.05976	.06093	.06210	.00023	.000063
109	.05843	.05960	.06076	.06193	.00023	.000064
110	.05827	.05943	.06060	.06176	.00023	.000065

PERMISSION KAHOE AIR BALANCE COMPANY

DENSITY - 1

WEIGHT OF SATURATED AND PARTLY SATURATED AIR

Dry Bulb Temp. °F t	Barometric Pressure in Inches of Mercury				Increase in weight per .01" rise in barometer	Density increase per degree W.B. depression
	28	28.5	29	29.5		
108	.06326	.06443	.06560	.06677	.00023	.000063
109	.06309	.06426	.06543	.06659	.00023	.000064
110	.06293	.06409	.06525	.06642	.00023	.000065

PERMISSION KAHOE AIR BALANCE COMPANY

DENSITY - 1

WEIGHT OF SATURATED AND PARTLY SATURATED AIR

Dry Bulb Temp. °F t	Barometric Pressure in Inches of Mercury			Increase in weight per .01" rise in barometer	Density increase per degree W.B. depression
	30	30.5	31		
108	.06794	.06911	.07028	.00023	.000063
109	.06776	.06893	.07009	.00023	.000064
110	.06758	.06875	.06991	.00023	.000065

PERMISSION KAHOE AIR BALANCE COMPANY

ENTHALPY OF AIR
BTUH/CFM
BAROMETRIC PRESSURE
INCHES OF HG.

W.B. Temp.	20.00	20.50	21.00	21.50	22.00	22.50	23.00
40	57.10	57.86	58.60	59.38	60.13	60.88	61.65
41	58.89	59.68	60.43	61.21	61.99	62.75	63.55
42	60.67	61.48	62.25	63.05	63.84	64.62	65.44
43	62.45	63.27	64.06	64.89	65.70	66.50	67.32
44	64.21	65.06	65.87	66.70	67.54	68.35	69.19
45	65.98	66.84	67.67	68.52	69.37	70.20	71.06
46	67.95	68.82	69.68	70.54	71.41	72.26	73.13
47	69.92	70.81	71.68	71.10	72.00	72.87	73.74
48	71.87	72.78	73.67	71.30	75.46	76.36	77.24
49	73.82	74.74	75.66	76.56	77.48	78.39	79.29
50	75.77	76.71	77.64	78.56	79.47	80.43	81.34
51	77.95	78.89	79.85	80.79	81.72	82.69	83.60
52	80.11	81.07	82.05	83.00	83.95	84.94	85.87
53	82.29	83.27	84.24	85.21	86.17	87.18	88.12
54	84.44	85.42	86.42	87.41	88.40	89.41	90.37
55	86.60	87.59	88.59	89.59	90.60	91.62	92.61
56	89.01	90.02	91.05	92.06	93.09	94.13	93.67
57	91.42	92.44	93.49	94.52	95.55	96.62	97.63
58	93.83	94.87	95.93	96.98	98.02	99.09	100.13
59	96.21	97.27	98.35	99.42	100.49	101.57	102.63
60	98.60	99.67	100.76	101.85	102.94	104.02	105.11
61	101.28	102.00	103.50	104.57	105.69	106.78	107.89
62	103.96	104.28	106.22	107.33	108.44	109.55	110.65
63	106.67	106.58	108.90	110.02	111.13	112.28	113.45
64	109.32	108.84	111.58	112.76	113.87	115.04	116.19

PERMISSION KAHOE AIR BALANCE COMPANY

ENTHALPY OF AIR
BTUH/CFM
BAROMETRIC PRESSURE
INCHES OF HG.

W.B. Temp.	20.00	20.50	21.00	21.50	22.00	22.50	23.00
65	111.96	111.13	114.27	115.45	116.58	117.76	118.93
66	115.00	114.57	117.32	118.50	119.69	120.86	122.03
67	118.01	118.00	120.36	121.55	122.73	123.94	125.12
68	121.05	121.43	123.43	124.63	125.82	126.98	128.23
69	124.03	124.83	126.44	127.66	128.83	130.06	131.29
70	127.02	128.21	129.44	130.67	131.89	133.10	134.34
71	130.42	131.61	132.87	134.10	135.34	136.57	137.82
72	133.81	135.03	136.26	137.51	138.77	140.02	141.26
73	137.15	138.38	139.66	140.93	142.16	143.45	144.71
74	140.50	141.76	143.04	144.31	145.56	146.85	148.14
75	143.84	145.10	146.38	147.68	148.94	150.25	151.56
76	147.68	148.95	150.24	151.57	152.84	154.15	155.42
77	151.50	152.78	154.08	155.39	156.68	157.99	159.32
78	155.30	156.60	157.90	159.25	160.55	161.87	163.17
79	159.08	160.39	161.71	163.02	164.39	165.68	167.04
80	162.85	164.17	165.49	166.85	168.19	169.51	170.86
81	167.19	168.52	169.83	171.21	172.48	173.85	175.21
82	171.45	172.81	174.14	175.54	176.73	178.21	179.57
83	175.76	177.08	178.44	179.82	180.96	182.55	183.88
84	180.01	181.36	182.75	184.11	185.19	186.87	188.24
85	184.24	185.59	187.00	188.38	189.38	191.13	192.51
86	189.17	190.50	191.92	193.31	194.40	196.11	197.48
87	194.08	195.42	196.82	198.25	199.41	201.00	202.42
88	198.95	200.36	201.76	203.12	204.38	205.94	207.34
89	203.81	205.19	206.61	208.01	209.34	210.82	212.24

PERMISSION KAHOE AIR BALANCE COMPANY

ENTHALPY OF AIR
BTUH/CFM
BAROMETRIC PRESSURE
INCHES OF HG.

W.B. Temp.	20.00	20.50	21.00	21.50	22.00	22.50	23.00
90	208.65	210.05	211.48	212.84	214.27	215.71	217.12
91	214.19	215.55	216.98	218.37	219.82	221.22	222.68
92	219.70	221.07	222.49	223.91	225.34	226.76	228.18
93	225.22	226.60	227.98	229.42	230.83	232.26	233.72
94	230.67	232.06	233.43	234.86	236.29	237.73	239.20
95	236.09	237.44	238.86	240.32	241.72	243.18	244.61
96	242.43	243.78	245.20	246.63	248.04	249.49	250.91
97	248.74	250.10	251.45	252.95	254.37	255.78	257.22
98	254.98	256.34	257.76	259.16	260.58	262.03	263.47
99	261.27	262.58	264.00	265.41	266.80	268.25	269.72
100	267.48	268.80	270.21	271.59	273.04	274.43	275.90

PERMISSION KAHOE AIR BALANCE COMPANY

ENTHALPY OF AIR
BTUH/CFM
BAROMETRIC PRESSURE
INCHES OF HG.

W.B. Temp.	23.50	24.00	24.50	25.00	25.50	26.00	26.50
40	62.44	63.20	63.92	64.68	65.46	66.22	67.00
41	64.32	65.13	65.86	66.66	67.43	68.23	68.99
42	66.25	67.05	67.82	68.63	69.39	70.23	70.99
43	68.12	68.92	69.74	70.53	71.39	72.17	73.03
44	70.03	70.84	71.78	72.49	73.34	74.17	75.01
45	71.90	72.74	73.61	74.43	75.27	76.15	76.98
46	73.99	74.85	75.71	76.58	77.43	78.29	79.18
47	76.07	76.93	77.84	78.72	79.57	80.48	81.36
48	78.15	79.06	79.92	80.81	81.75	82.61	83.54
49	80.22	81.13	82.05	82.93	83.87	84.78	85.70
50	82.29	83.20	84.12	85.04	85.98	86.90	87.86
51	84.56	85.49	86.42	87.40	88.35	89.26	90.23
52	86.85	87.79	88.76	89.75	90.69	91.61	92.63
53	89.12	90.10	91.04	92.04	93.03	94.01	94.98
54	91.36	92.36	93.37	94.37	95.37	96.34	97.36
55	93.62	94.63	95.63	96.68	97.70	98.66	99.69
56	96.14	97.15	98.17	99.24	100.25	101.29	102.33
57	98.68	99.72	100.74	101.83	102.86	103.91	104.94
58	101.18	102.23	103.28	104.37	105.41	106.45	107.50
59	103.69	104.76	105.83	106.93	107.99	109.05	110.11
60	106.16	107.26	108.34	109.45	110.52	111.64	112.70
61	108.98	110.08	111.19	112.28	113.38	114.50	115.57
62	111.78	112.91	113.99	115.14	116.25	117.34	118.48
63	114.55	115.67	116.82	117.95	119.08	120.24	121.35
64	117.33	118.46	119.59	120.79	121.92	123.06	124.24

PERMISSION KAHOE AIR BALANCE COMPANY

ENTHALPY OF AIR
BTUH/CFM
BAROMETRIC PRESSURE
INCHES OF HG.

W.B. Temp.	23.50	24.00	24.50	25.00	25.50	26.00	26.50
65	120.07	121.25	122.39	123.58	124.73	125.89	127.07
66	123.21	124.37	125.56	126.76	127.90	129.11	130.27
67	126.30	127.49	128.70	129.89	131.06	132.25	133.48
68	129.41	130.64	131.78	133.04	134.24	135.41	136.62
69	132.47	133.72	134.91	136.14	137.38	138.57	139.81
70	135.56	136.80	138.01	139.27	140.49	141.71	142.96
71	139.03	140.28	141.53	142.79	144.03	145.25	146.53
72	142.48	143.77	145.03	146.30	147.55	148.81	150.08
73	145.98	147.22	148.48	149.80	151.04	152.31	153.61
74	149.40	150.69	151.95	153.28	154.53	155.86	157.14
75	152.83	154.12	155.41	156.74	158.03	159.33	160.64
76	156.71	158.02	159.36	160.66	161.98	163.28	164.60
77	160.64	161.94	163.26	164.61	165.93	167.25	168.58
78	164.48	165.83	167.18	168.47	169.83	171.17	172.53
79	168.37	169.71	171.04	172.39	173.76	175.10	176.48
80	172.21	173.57	174.95	176.27	177.64	178.98	180.37
81	176.60	177.92	179.33	180.68	182.07	183.43	184.83
82	180.96	182.32	183.72	185.07	186.47	187.86	189.24
83	185.31	186.66	188.06	189.43	190.85	192.22	193.62
84	189.64	191.02	192.41	193.81	195.22	196.61	198.03
85	193.94	195.32	196.71	198.13	199.56	200.97	202.38
86	198.89	200.29	201.71	203.13	204.57	205.97	207.40
87	203.85	205.22	206.69	208.08	209.51	210.95	212.41
88	208.75	210.17	211.61	213.08	214.51	215.95	217.35
89	213.71	215.10	216.51	217.98	219.41	220.88	222.31

PERMISSION KAHOE AIR BALANCE COMPANY

ENTHALPY OF AIR
BTUH/CFM
BAROMETRIC PRESSURE
INCHES OF HG.

W.B. Temp.	23.50	24.00	24.50	25.00	25.50	26.00	26.50
90	218.56	220.00	221.42	222.90	224.36	225.79	227.25
91	224.12	225.56	226.99	228.49	229.93	231.40	232.84
92	229.65	231.08	232.54	234.02	235.51	236.95	238.39
93	235.15	236.61	238.02	239.51	240.98	242.47	243.96
94	240.63	242.08	243.51	245.03	246.51	248.01	249.46
95	246.08	247.56	248.98	250.42	251.96	253.48	254.97
96	252.41	253.85	255.28	256.81	258.30	259.80	261.27
97	258.65	260.12	261.63	263.12	264.60	266.09	267.58
98	264.92	266.43	267.87	269.36	270.88	272.35	273.87
99	271.15	272.63	274.12	275.61	277.12	278.58	280.12
100	277.30	278.85	280.34	281.79	283.34	284.78	286.35

PERMISSION KAHOE AIR BALANCE COMPANY

ENTHALPY OF AIR
BTUH/CFM
BAROMETRIC PRESSURE
INCHES OF HG.

W.B. Temp.	27.00	27.50	28.00	28.50	29.00	29.50	30.00
40	67.75	68.49	69.27	70.02	70.80	71.57	72.33
41	69.80	70.53	71.34	72.08	72.91	73.71	74.46
42	71.83	72.61	73.41	74.19	75.00	75.80	76.57
43	73.81	74.63	75.42	76.23	77.08	77.92	78.73
44	75.84	76.69	77.46	78.32	79.16	79.98	80.83
45	77.85	78.69	79.51	80.36	81.24	82.09	82.92
46	80.04	80.92	81.74	82.62	83.48	84.39	85.21
47	82.22	83.15	83.99	84.88	85.73	86.62	87.54
48	84.44	85.31	86.21	87.09	88.03	88.91	81.46
49	86.60	87.51	88.46	89.32	90.26	91.13	92.12
50	88.76	89.71	90.64	91.55	92.48	93.40	94.38
51	91.17	92.12	93.05	94.01	94.94	95.89	96.87
52	93.57	94.53	95.51	96.46	97.43	98.38	99.35
53	95.92	96.98	97.90	98.89	99.87	100.85	101.88
54	98.31	99.35	100.33	101.33	102.34	103.33	104.33
55	100.68	101.74	102.71	103.75	104.76	105.79	106.80
56	103.32	104.38	105.41	106.44	107.46	108.48	109.53
57	105.95	107.02	108.04	109.14	110.14	111.21	112.27
58	108.62	109.69	110.72	111.76	112.86	113.89	114.94
59	111.22	112.30	113.34	114.42	115.53	116.61	117.65
60	113.82	114.90	115.98	117.08	118.20	119.26	120.36
61	116.71	117.79	118.95	120.04	121.15	122.25	123.34
62	119.59	120.74	121.84	122.95	124.12	125.20	126.32
63	122.49	123.61	124.77	125.89	127.02	128.16	129.32
64	125.35	126.52	127.65	128.77	129.96	131.11	132.25

PERMISSION KAHOE AIR BALANCE COMPANY

ENTHALPY OF AIR
BTUH/CFM
BAROMETRIC PRESSURE
INCHES OF HG.

W.B. Temp.	27.00	27.50	28.00	28.50	29.00	29.50	30.00
65	128.19	129.38	130.56	131.70	132.89	134.05	135.20
66	131.42	132.64	133.83	134.98	136.18	137.34	138.52
67	134.66	135.83	137.05	138.24	139.41	140.63	141.82
68	137.82	139.04	140.28	141.46	142.70	143.87	145.11
69	141.01	142.21	143.47	144.70	145.90	147.14	148.40
70	144.19	145.43	146.69	147.94	149.14	150.39	151.67
71	147.79	149.00	150.28	151.56	152.79	154.07	155.29
72	151.33	152.59	153.86	155.17	156.43	157.65	158.95
73	154.88	156.15	157.43	158.72	159.97	161.30	162.57
74	158.39	159.71	161.00	162.29	163.58	164.87	166.19
75	161.93	163.24	164.53	165.85	167.16	168.46	173.14
76	165.93	167.24	168.57	169.86	171.20	167.80	176.56
77	169.91	171.22	172.56	173.91	175.22	176.59	179.91
78	173.84	175.20	176.52	177.88	179.22	180.57	183.28
79	177.79	179.15	180.48	181.88	183.98	184.59	186.61
80	181.72	183.08	184.45	185.84	187.20	188.59	189.97
81	186.16	187.53	188.91	190.34	191.68	193.12	194.48
82	190.62	191.96	193.39	194.80	196.17	197.57	198.98
83	195.02	196.42	197.81	199.22	200.65	202.05	203.50
84	199.44	200.84	202.25	203.67	205.10	206.49	207.96
85	203.80	205.22	206.63	208.10	209.49	210.93	212.39
86	208.85	210.26	211.70	213.13	214.59	216.04	217.47
87	213.81	215.27	216.69	218.16	219.61	221.08	222.55
88	218.82	220.27	221.71	223.18	224.66	226.10	227.59
89	223.77	225.24	226.70	228.19	229.64	231.10	232.60

PERMISSION KAHOE AIR BALANCE COMPANY

ENTHALPY OF AIR
BTUH/CFM
BAROMETRIC PRESSURE
INCHES OF HG.

W.B. Temp.	27.00	27.50	28.00	28.50	29.00	29.50	30.00
90	228.73	230.19	231.67	233.14	234.61	236.11	237.62
91	234.36	235.81	237.29	238.74	240.24	241.76	243.25
92	239.87	241.40	242.87	244.36	245.83	247.38	248.89
93	245.44	246.93	248.43	249.90	251.45	252.94	254.42
94	250.94	252.46	253.95	255.50	257.00	258.51	260.01
95	256.45	257.97	259.46	261.00	262.55	264.05	265.58
96	262.77	264.31	265.80	267.34	268.90	270.39	271.95
97	269.06	270.61	272.15	273.65	275.21	276.74	278.25
98	275.40	276.92	278.39	279.94	281.45	283.02	284.56
99	281.63	283.17	284.68	286.19	287.75	289.35	290.85
100	287.83	289.38	290.90	292.46	293.97	295.57	297.10

PERMISSION KAHOE AIR BALANCE COMPANY

ENTHALPY OF AIR
BTUH/CFM
BAROMETRIC PRESSURE
INCHES OF HG.

W.B. Temp.	30.50	31.00	W.B. Temp.	30.50	31.00
40	73.09	73.84	70	152.88	154.14
41	75.23	76.15	71	156.57	157.80
42	77.38	78.18	72	160.20	161.50
43	79.53	80.31	73	163.86	165.13
44	81.67	82.48	74	167.46	168.80
45	83.77	84.59	75	171.09	172.40
46	86.07	86.96	76	174.59	175.89
47	88.39	89.27	77	178.04	179.36
48	90.69	91.61	78	181.50	182.87
49	93.05	93.95	79	184.93	186.26
50	95.29	96.23	80	188.36	189.76
51	97.78	98.76	81	193.48	194.87
52	100.25	101.28	82	198.60	200.04
53	102.66	103.83	83	203.70	205.11
54	105.11	106.33	84	208.77	210.23
55	107.56	108.81	85	213.83	215.28
56	110.36	111.57	86	218.90	220.38
57	113.15	114.32	87	224.01	225.44
58	115.95	117.10	88	229.05	230.52
59	118.72	119.81	89	234.06	235.53
60	121.48	122.55	90	239.05	240.57
61	124.48	125.58	91	244.73	246.23
62	127.48	128.58	92	250.35	251.87
63	130.44	131.57	93	255.95	257.48
64	133.43	134.57	94	261.55	263.10

PERMISSION KAHOE AIR BALANCE COMPANY

ENTHALPY OF AIR
BTUH/CFM
BAROMETRIC PRESSURE
INCHES OF HG.

W.B. Temp.	30.50	31.00	W.B. Temp.	30.50	31.00
65	136.36	137.56	95	267.10	268.65
66	139.72	140.89	96	273.50	275.00
67	143.00	144.20	97	279.83	281.35
68	146.31	147.56	98	286.13	287.69
69	149.59	150.87	99	292.40	293.98
			100	298.69	300.21

PERMISSION KAHOE AIR BALANCE COMPANY

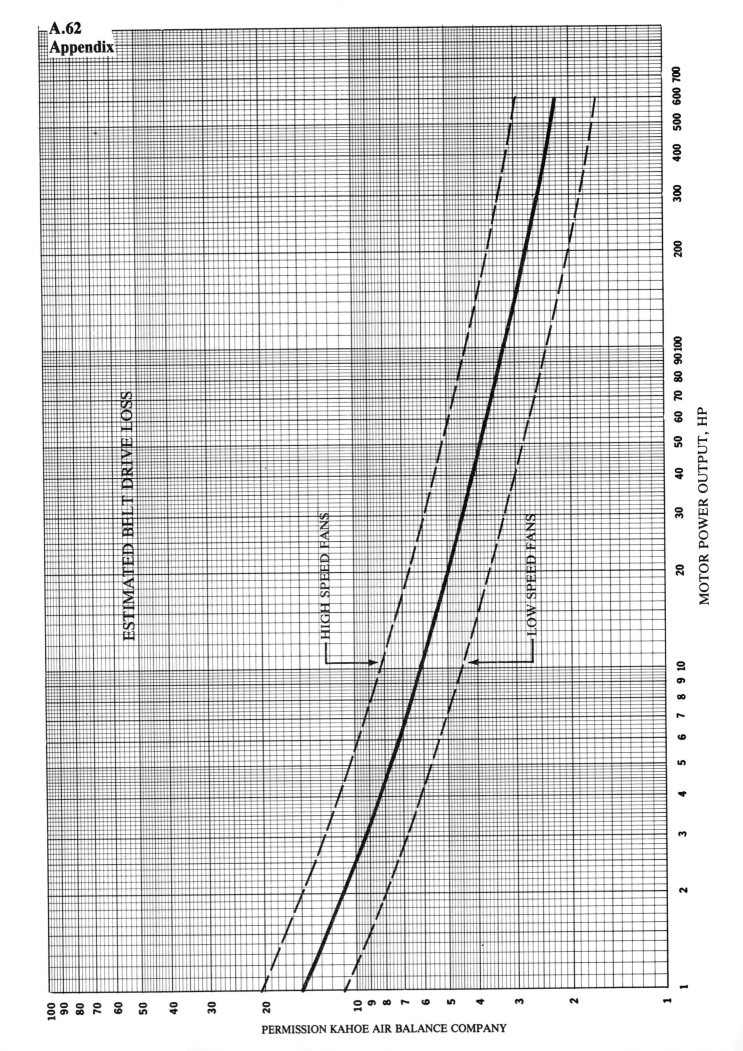

FAN LAW 1 — CONSTANT SYSTEM — SPEED VARIABLE

$$\frac{CFM_1}{CFM_2} = \frac{RPM_1}{RPM_2}$$

CFM varies directly as the RPM.

$$\frac{SP_1}{SP_2} = \left(\frac{RPM_1}{RPM_2}\right)^2$$

Static Pressure varies as the square of the RPM.

$$\frac{BHP_1}{BHP_2} = \left(\frac{RPM_1}{RPM_2}\right)^3$$

Brake Horsepower varies as the cube of the RPM.

FAN LAW 2 — CONSTANT TIP SPEED, AIR DENSITY, FAN PROPORTIONS, AND FIXED OPERATING POINT — FAN SIZE VARIABLE.

$$\frac{CFM_1}{CFM_2} = \frac{BHP_1}{BHP_2} = \left(\frac{D_1}{D_2}\right)^2$$

CFM and BHP varies as the Square of wheel diameter.

$$SP_1 = SP_2$$

Static pressure remains constant.

$$\frac{RPM_1}{RPM_2} = \frac{D_2}{D_1}$$

RPM varies inversely as the wheel diameter.

PERMISSION KAHOE AIR BALANCE COMPANY

FAN LAW 3 — CONSTANT RPM, AIR DENSITY, FAN PROPORTIONS AND FIXED OPERATING POINT. FAN SIZE VARIES.

$$\frac{CFM_1}{CFM_2} = \left(\frac{D_1}{D_2}\right)^3 \quad \text{CFM varies as the cube of the wheel diameter.}$$

$$\frac{SP_1}{SP_2} = \left(\frac{D_1}{D_2}\right)^2 \quad \text{Static pressure varies as the square of the wheel diameter.}$$

$$\frac{TS_1}{TS_2} = \frac{D_1}{D_2} \quad \text{Tip speed varies directly as the wheel diameter.}$$

$$\frac{BHP_1}{BHP_2} = \left(\frac{D_1}{D_2}\right)^5 \quad \text{Brake horsepower varies directly as the fifth power of wheel diameter.}$$

FAN LAW 4 — CONSTANT VOLUME, SYSTEM, FAN SIZE AND RPM – AIR DENSITY CHANGE.

$$CFM_1 = CFM_2 \quad \text{Air volume remains constant}$$

$$\frac{SP_1}{SP_2} = \frac{BHP_1}{BHP_2} = \frac{d_1}{d_2} \quad \text{Static pressure and BHP varies in direct proportion to the density.}$$

PERMISSION KAHOE AIR BALANCE COMPANY

FAN LAW 5   CONSTANT PRESSURE, SYSTEM AND FAN SIZE – RPM CHANGE.

$$\frac{CFM_1}{CFM_2} = \frac{RPM_1}{RPM_2} = \frac{BHP_1}{BHP_2} = \left(\frac{d_2}{d_1}\right)^{1/2}$$

CFM, RPM, and BHP vary inversely as the square root of the density.

$$S.P._1 = S.P._2 \qquad \text{Static pressure remains constant.}$$

FAN LAW 6   CONSTANT MASS FLOW RATE, CONSTANT SYSTEM AND FIXED FAN SIZE – RPM CHANGE.

$$\frac{CFM_1}{CFM_2} = \frac{RPM_1}{RPM_2} = \frac{SP_1}{SP_2} = \frac{d_2}{d_1} \qquad \text{CFM, RPM and SP varies inversely with the density.}$$

$$\frac{BHP_1}{BHP_2} = \left(\frac{d_2}{d_1}\right)^2 \qquad \text{BHP varies inversely with the square of the density.}$$

PERMISSION KAHOE AIR BALANCE COMPANY

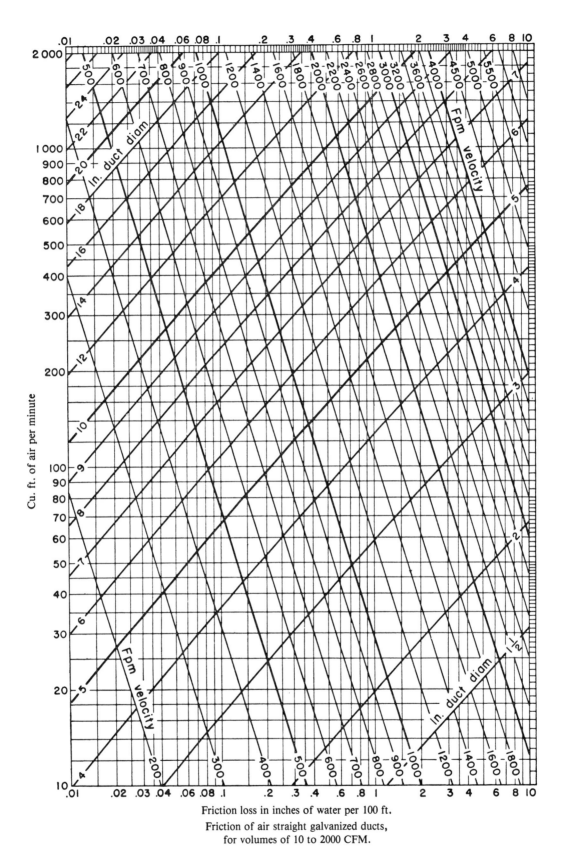

Friction of air straight galvanized ducts, for volumes of 10 to 2000 CFM.

(Based on standard air of 0.075 lb. per cu. ft. density flowing through clean round galvanized ducts having approximately 40 joints per 100 ft.

PERMISSION KAHOE AIR BALANCE COMPANY

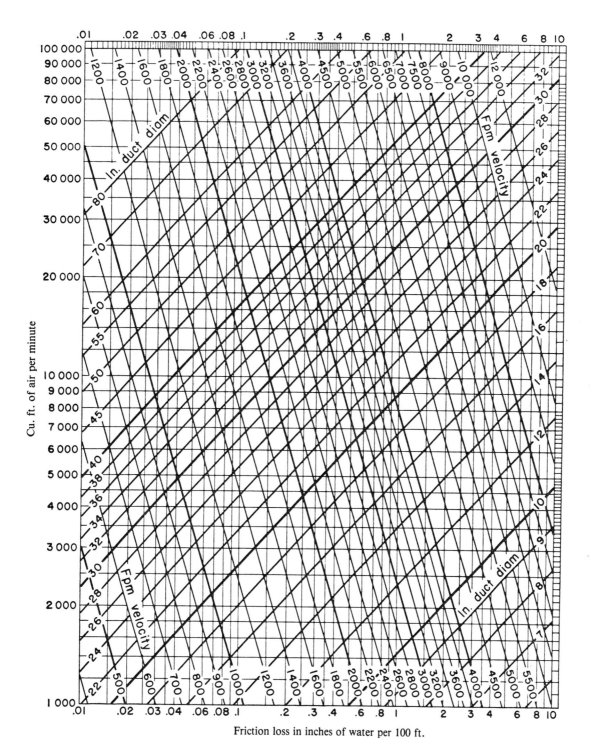

Friction of air in straight galvanized ducts, for volumes of 1000 to 100,000 CFM.

PERMISSION KAHOE AIR BALANCE COMPANY

A.68
Appendix

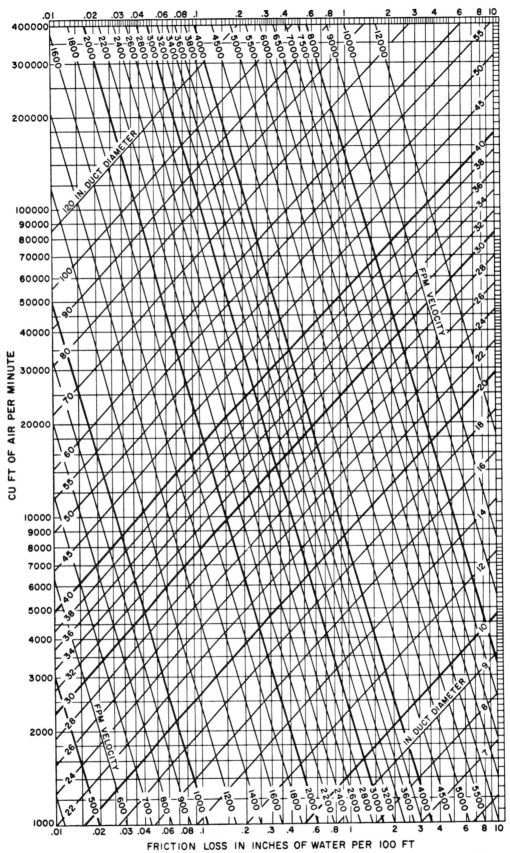

(Based on Standard Air of 0.075 lb per cu ft density flowing through average, clean, round, galvanized metal ducts having approximately 40 joints per 100 ft.)

Friction of Air in Straight Ducts for Volumes of 1000 to 400,000 Cfm

PERMISSION KAHOE AIR BALANCE COMPANY

HP	AVERAGE FULL-LOAD CURRENTS OF THREE-PHASE MOTORS							
	Full load currents for squirrel-cage and wound-rotor induction types					Full-load currents for synchronous unity-power-factor type			
	110 volts	220 volts	440 volts	550 volts	2300 volts	220 volts	440 volts	550 volts	2300 volts
1/2	4	2	1	0.8					
3/4	5.6	2.8	1.4	1.1					
1	7	3.5	1.8	1.4					
1-1/2	10	5	2.5	2.0					
2	13	6.5	3.3	2.6					
3	---	9	4.5	4					
5	---	15	7.5	6					
7-1/2	---	22	11	9					
10	---	27	14	11					
15	---	40	20	16					
20	---	52	26	21					
25	---	64	32	26	7	54	27	22	5.4
30	---	78	39	31	8.5	65	33	26	6.5
40	---	104	52	41	10.5	86	43	35	8
50	---	125	63	50	13	108	54	44	10
60	---	150	75	60	15	128	64	51	12
75	---	185	93	74	19	161	81	65	15
100	---	246	123	98	25	211	106	85	20
125	---	310	155	124	31	264	132	106	25
150	---	360	180	144	37	---	158	127	30
200	---	480	240	192	49	---	210	168	40

PERMISSION KAHOE AIR BALANCE COMPANY

AVERAGE FULL-LOAD CURRENTS OF D-C MOTORS

HP	115 volts	230 volts	550 volts	HP	115 volts	230 volts	550 volts
	Full-load current amp.				Full-load current amp.		
1/2	4.6	2.3		20	148	74	31
3/4	6.6	3.3	1.4	25	184	92	38
1	8.6	4.3	1.8	30	220	110	46
1-1/2	12.6	6.3	2.6	40	292	146	61
2	16.4	8.2	3.4	50	360	180	75
3	24.0	12.0	5.0	60	430	215	90
5	40	20	8.3	75	536	268	111
7-1/2	58	29	12.0	100	---	355	148
10	76	38	16.0	125	---	443	184
15	112	56	23.0	150	---	534	220
				200	---	712	295

AVERAGE FULL-LOAD CURRENTS OF SINGLE-PHASE MOTORS

HP	110 volts	220 volts	440 volts	HP	110 volts	220 volts	440 volts
	Full-load current amp.				Full-load current amp.		
1/6	3.2	1.6		1-1/2	18.4	9.2	
1/4	4.6	2.3		2	24	12	
1/2	7.4	3.7		3	34	17	
3/4	10.2	5.1		5	56	28	
1	13	6.5		7-1/2	80	40	21
				10	100	50	26

PERMISSION KAHOE AIR BALANCE COMPANY

APPROXIMATE P.F. AND EFF. OF SQUIRREL CAGE INDUCTION MOTORS (U-FRAME)					
	1/2 Load		3/4 Load		Full Load	
HP	P.F.	Eff.	P.F.	Eff.	P.F.	Eff.
1/2	44.8	65.1	56.2	71.1	69.2	71.6
3/4	49.0	69.5	60.0	73.7	72.0	74.7
1	54.3	75.1	67.2	78.5	76.5	79.2
1.5	59.2	76.3	72.2	79.3	80.5	80.3
2	64.3	79.7	76.7	82.2	85.3	82.3
3	64.3	79.8	79.5	82.0	82.6	81.6
5	69.5	82.8	79.3	84.0	84.2	83.5
7.5	72.7	84.8	80.4	86.0	85.5	85.5
10	75.8	83.3	80.3	85.5	88.8	85.5
15	77.8	86.0	81.4	87.3	87.0	87.3
20	78.3	88.3	83.2	89.3	87.2	88.7
25	75.8	87.8	82.8	89.2	86.8	88.7
30	77.3	88.7	83.5	89.7	87.2	89.7
40	80.0	90.0	85.7	90.8	88.2	90.7
50	80.2	89.3	85.7	90.3	89.2	90.0
60	81.8	89.1	86.3	90.5	89.5	90.2
75	83.2	90.3	87.2	91.3	89.5	91.2
100	86.3	90.3	89.5	91.3	90.3	91.2
125	85.3	90.3	89.3	91.3	90.5	91.5
150	85.0	90.3	88.8	91.8	90.5	92.0

NOTE: This chart can be used to determine approximate efficiency where power factor has been taken on these size motors.

PERMISSION KAHOE AIR BALANCE COMPANY

Appendix

AIR VOLUME BY PRESSURE DROP ACROSS A SQUARE EDGE ORIFICE

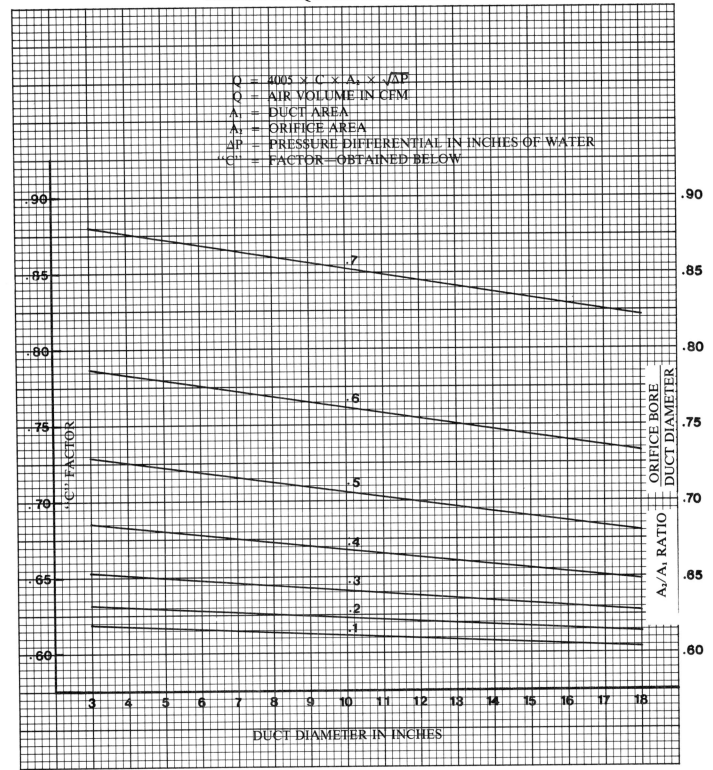

PERMISSION KAHOE AIR BALANCE COMPANY

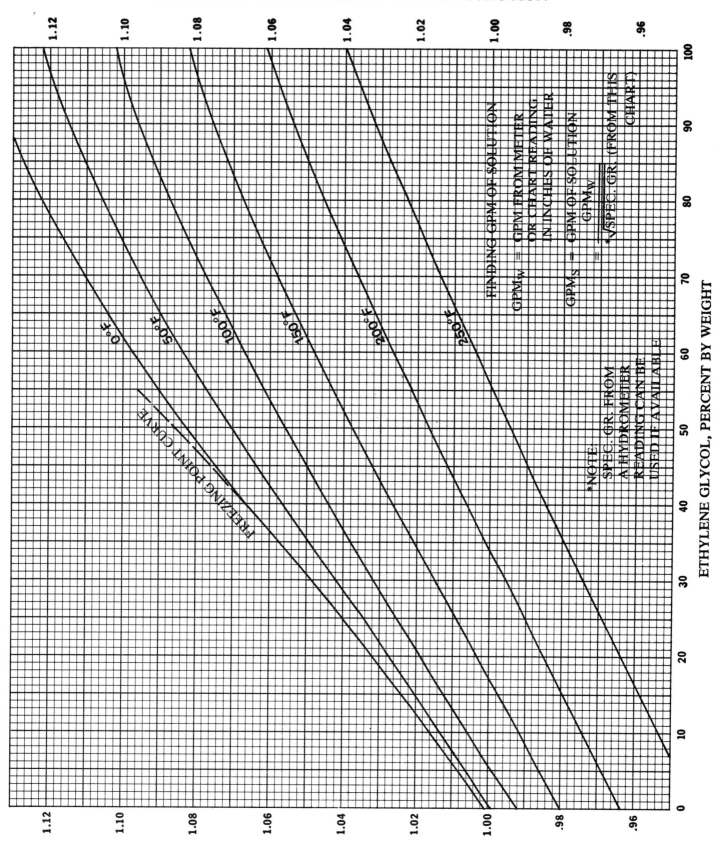